Elementary Textbook of Anatomy and Physiology Applied to Nursing

BY

Janet T. E. Riddle
R.G.N., R.F.N., O.N.C.

Registered Nurse Tutor (Edin.). Member of the
Panel of Examiners for the Roll of the General Nursing Council for Scotland.
Examiner to the Joint Examination Board, British Orthopaedic Association and
Central Council for the Disabled.
Member of the Joint Examination Board.

ILLUSTRATED BY
Kathleen B. Nicoll
R.G.N., S.C.M., O.N.C., R.C.I.(Edin.), R.N.T.

FOURTH EDITION

Churchill Livingstone   Edinburgh & London 1974

CHURCHILL LIVINGSTONE
Medical Division of Longman Group Limited

Represented in the United States of America by Longman Inc., New York and by associated companies, branches and representatives throughout the world.

© E. & S. Livingstone Limited, 1961, 1966
© Longman Group Limited, 1969, 1974

All rights reserved. No part of this publication may be reproduced, stored in a retrieval system, or transmitted in any form or by any means, electronic, mechanical, photocopying, recording or otherwise, without the prior permission of the publishers (Churchill Livingstone, 23 Ravelston Terrace, Edinburgh).

First Edition 1961
Reprinted 1963
Second Edition 1966
Reprinted 1967
Reprinted 1968
Third Edition 1969
Reprinted 1970
Reprinted 1972
Fourth Edition 1974

ISBN 0 443 01097 8

Library of Congress Catalog Card Number 73-85675

PRINTED IN GREAT BRITAIN

Preface to Fourth Edition

In the preparation of this fourth edition, I have had helpful advice from many of my colleagues especially Miss Elizabeth Dunsire and Miss Alison I. Murchie. As a result I have made several alterations mainly of an explanatory nature and have added a section on blood groups. Miss Nicoll has produced one more illustration.

I am grateful for the help of all my colleagues and indebted to Mrs Helen Watson for typing the revised script.

Janet T. E. Riddle.

1974

Preface to First Edition

This book is based on the lectures given to the nurses at Killearn Hospital during the first year of their training. During this year an attempt is made to give the student a simple overall picture of the human body and to make her apply her knowledge to the art of Nursing which she is learning on the wards and in the classroom. The nurse, training for State Enrolment, requires no more than an elementary knowledge of Anatomy and Physiology. If these subjects are taught without practical application, they become the tedious and dreaded stumbling blocks to the passing of examinations. Applied to nursing, however, Anatomy and Physiology become interesting and alive and much that was bewildering becomes obvious.

The chapter on Posture has been placed at the end of the book for ease of reference although its place is more rightly following Chapter 3. This is a subject which cannot be disposed of in one lecture, but must be taught over and over again particularly at the bedside.

I wish to express my grateful thanks to all who have helped in the preparation of this book; particularly Miss K. B. Nicoll who not only prepared the illustrations but typed the manuscript and assisted in the preparation of the material; Mr. Athol Parkes for reading the manuscript; Professor Roland Barnes and Miss E. Macinnes for their helpful criticisms; Mr. Charles Macmillan, Mr. James Parker and the staff of E. & S. Livingstone Ltd. for the encouragement, advice and guidance which they have given so willingly. Many other friends as well as members of my family have contributed their help and encouragement and I am greatly indebted to them all.

JANET T. E. RIDDLE.

Killearn, 1961

Contents

Preface
1. The Structure of the Body — 1
2. The Skeletal System — 9
3. The Joints and Muscles — 42
4. The Circulatory System — 53
5. The Respiratory System — 69
6. The Digestive System — 77
7. The Excretory Systems — 95
8. The Nervous System and the Special Senses — 105
9. The Endocrine or Ductless Glands — 119
10. The Reproductive Systems — 127
11. Posture—Nurse and Patient — 133
    Index — 144

1. The Structure of the Body

Biology is the study of life. This book is about **human** life. In the study of **Anatomy** we are concerned with our own structure. **Physiology** tells us how our bodies function. We are like very delicate machines, every part beautifully constructed to carry out its own particular function but also to work as a whole with all the other parts.

You can compare the human body with a machine; we will be doing it again and again in the following chapters, but you must also remember that, unlike machines, we have souls. No doctor or nurse must ever become like a mechanic repairing or servicing the body without any thought for the person. Although we are all constructed on similar lines, each one of us has spiritual, emotional and intellectual qualities of varying degrees. So often we hear of and meet nurses who are excellent mechanics. They nurse and tend the broken down body expertly but have no interest in and no concern for the individual.

The purpose of this book is not to try to define the soul but to describe the body it inhabits. As a nurse you must learn how the body works, how to help repair the broken parts and what to do to prevent it from breaking down, but, as a good nurse you must also learn to provide for the spiritual and emotional needs of your patients.

Cells and Tissues

The human body can be described as a multi-cellular animal because it is built up of billions of microscopic specks of living material called **cells** (Fig. 1.1). These cells contain a very complicated substance called protoplasm and each cell consists of a ball of protoplasm surrounded by a cell wall with a central nucleus. The nucleus is the part of the cell responsible for its life and contains minute specks of material which pass on characteristics from one generation to the next. The fact that you and your grandmother have red hair is not just coincidence.

The cell wall is semi-permeable allowing water and substances in solution to pass in and out of the protoplasm. This is important because you breathe so that each cell can get oxygen and you eat so that each cell can get the food substances it needs.

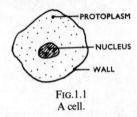

FIG.1.1
A cell.

Each cell is a living thing and in some way or other has all the **characteristics of life.** Each cell can **grow** and **repair** itself, **move, breathe, take in food, get rid of waste, reproduce** itself and **react** to its environment.

The cells are stuck together in various ways to form **tissues** (Fig. 1.2). These tissues are really the materials from which the body is made and each tissue has its own special function.

There are four different types of tissue, **epithelial tissue, connective tissue, muscle tissue,** and **nervous tissue.** Specimens from the butcher, for example a joint and a sheep's heart, will show you most of the tissues:

EPITHELIAL TISSUE (EPITHELIUM)

There are various types of epithelium. These are mainly lining tissues and you will see a very fine type lining the heart. This very delicate tissue is also found where substances in solution must pass from one part to another. It is semi-permeable and is found in the lungs.

A thicker type of epithelium lines the whole of your digestive tract and is called **mucous membrane** because it secretes a sticky mucus. In the respiratory passages this mucous membrane has small hair like processes called cilia growing from the surface. The cilia help to trap the dust and filter the air we breathe. This tissue is called ciliated epithelium.

The thickest type of epithelial tissue forms the outer layer of the skin.

CONNECTIVE TISSUES

Examine the joint you have purchased from the butcher. **Bone** is a connective tissue. The **tendons** which attach the muscle to the bone

The Structure of the Body

and the **ligaments** which act like ropes joining the bones together are also connective tissues. These tissues have little thread like structures called fibres between the cells, this makes them strong. They are called **fibrous tissues**.

There are other fibrous tissues which form protective coverings. Notice the fine 'skin', rather like tissue paper, which is wrapped round the bone and the muscle. Examine the fatty tissue, you can see its fibrous structure and also that it contains a great deal of fat. This tissue is called **adipose tissue**. It is found under the skin preventing the loss of heat and around the kidneys giving them protection.

Other connective tissues are the cartilages. **Hyaline cartilage** is the bluish-white shiny substance covering the ends of the bones where they join. This cartilage is a tough material and will stand up to great stresses and strains. It is not brittle like bone and does not break easily.

MUSCLE TISSUE

Examine the muscle tissue attached to the bone. This is **voluntary** or **skeletal** muscle tissue. It is controlled by the brain. The other types of muscle tissue are **involuntary** forming the walls of the internal organs and **cardiac** forming the walls of the heart.

NERVOUS TISSUE

This tissue is soft **grey** or **white** in colour and forms the brain and spinal cord.

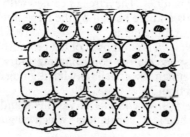

Fig. 1.2
Tissue.

Systems and Organs

The body is composed of nine different systems each one with its own function and each one co-operating with and helping the others. Here is true team spirit.

The **systems** are made up of a collection of **organs** and each organ is made of several **tissues**. For example, the heart is an organ of the circulatory system; it is made of muscle because its function is to pump; it is lined with smooth epithelium to give the blood a smooth surface to flow over and it is covered with tough fibrous tissue which protects it. It can be compared with a house built of brick because its function is to provide shelter, covered with rough-cast for protection and lined with smooth plaster.

The systems

The *Skeletal System* is made up of bones and its function is to provide a framework.

The *Muscular System* is composed of muscles which produce movement.

The *Circulatory System* consists of the heart, blood vessels, lymphatics and spleen and their function is to provide the cells of the body with nourishment and to carry away waste products.

The *Respiratory System* consists of the lungs and the air passages and its function is to take in air containing oxygen and to give out air containing carbon dioxide.

The *Digestive System* is made up of the alimentary canal which includes the mouth, the pharynx, the oesophagus, the stomach and the intestines, and the glands including the pancreas, liver and salivary glands. The function of this system is to ingest, digest, absorb food, and excrete waste.

The *Excretory System* consists of the skin, and the kidneys with their ureters and bladder. They are concerned with ridding the body of waste materials and are helped by the large intestine, part of the digestive system and the lungs.

The *Nervous System* is composed of the brain, spinal cord and the nerves and is responsible for co-ordinating the work of the other systems. It is the system which tells us what is going on around us and controls our reactions.

The *Endocrine System* consists of various glands which produce hormones or chemical substances which regulate the body's activities.

The *Reproductive System* consists of the ovaries, uterus and vagina in the female and the testes, prostate gland and penis in the male. It is concerned with the reproduction of the human being.

The Anatomical Parts of the Body

The body can be divided into the head, the neck, the trunk, and the limbs.

The head
This contains the **cranium,** a large cavity in which lies the brain. It also has four smaller cavities, the **nose,** the **orbits** which contain the eyes and the **mouth** in which lies the tongue.

The neck
This joins the head to the trunk and consists of part of the spinal column and the food and the air passages passing down into the trunk.

The trunk (Fig. 1.3)
The trunk is divided into three cavities:
    The abdomen.
    The thorax.
    The pelvis.

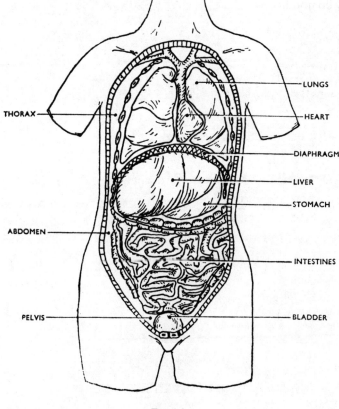

FIG. 1.3
The trunk.

The *Thorax* or chest is separated from the abdomen by a sheet of muscle called the **diaphragm.** This thoracic cavity contains the heart and lungs. These organs are protected by the ribs which form the walls of the thorax.

The *Abdomen* is the largest of the cavities and contains all the organs of the digestive system and also the kidneys. Its walls are mainly made of muscle.

The *Pelvic Cavity* is the lowest one. Its walls are of bone to protect the important reproductive organs as well as the bladder and rectum. The rectum is the end of the large intestine.

The limbs

The upper and lower limbs consist of bone covered by muscle and skin. Blood vessels and nerves run through these structures and their function is movement. They are not vital to life, many patients lose one or more limbs and although life is no longer so pleasant or useful they do not die.

Questions

Now try to answer these questions. You will find some questions at the end of each chapter. These are to help you to study and to test your knowledge. There are two types of questions, the recognition type where you recognize the correct answer amongst several incorrect answers and the recall type where you must recall what you have read and supply the correct answer or complete the statement as required.

(1) The study of the structure of the human body is ....
(2) .... is the study of the function of the parts of the body.
(3) The substance from which the cells are made is ....
(4) Inheritance depends on the structures contained within the .... of the cell.
(5) The characteristics of life are: (a) ....
   (b) .... (c) ....
   (d) .... (e) ....
   (f) .... (g) ....
(6) A collection of cells is: (a) an organ (b) a tissue (c) a system.
(7) Tendons are: (a) bone (b) fibrous tissue (c) cartilage (d) epithelium.
(8) A tendon is a structure which attaches .... to ....
(9) A .... is a structure which attaches bone to bone. It is also made of ....
(10) The tissue found lining organs is: (a) fat (b) cartilage (c) epithelium (d) muscle.
(11) Cartilage is: (a) harder than (b) not so hard as—bone.
(12) The type of cartilage found where the ends of bone join is called .... cartilage.
(13) One of the following organs belongs to the circulatory system: (a) the pancreas (b) the spleen (c) the lungs (d) the pharynx.
(14) One of the following is wrong. Which one? (a) the kidneys (b) the skin (c) the intestines (d) the liver—all excrete waste.
(15) Co-ordinating the work of all the systems of the body is the work of the .... system.
(16) A hormone is: (a) a part of the reproductive system (b) a chemical product of a gland (c) a digestive organ (d) a cell.
(17) The largest cavity of the trunk is the ....
(18) The diaphragm is a .... which separates the .... from the abdomen.
(19) The bladder is contained within the .... cavity.
(20) The stomach is contained within the .... cavity.

(21) Use this outline to indicate the position of the thorax, abdomen and pelvis, and draw in the organs using the classroom model as a guide.

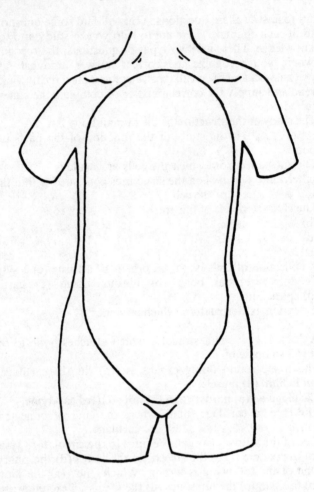

2. The Skeletal System

The skeleton is a framework. It is covered on the outside with muscle and skin and it forms the walls of the body cavities and the basis of the limbs.

Muscle is a remarkably healthy tissue, but skin and fibrous tissue are more easily affected by adverse circumstances. Where the bones are well covered by muscle they present no nursing problems. But where the bony prominences lie just under the skin they become important pressure points causing friction and pressure and limiting the blood supply to the skin over them.

It is obvious, therefore, that the nurse must have some knowledge of the structure of the skeleton in order to nurse her patient comfortably and also to make accurate reports on pressure areas.

The organs of the skeletal system are the bones, and, like all other organs, the bones are made up of several different tissues and are of different shapes.

If we take the shapes first it must be obvious from examining the skeleton and handling the bones that some are long or rod shape, some flat, while others are irregular in shape.

There are really four different types of bones:

Long bones
These form the limbs and allow us to move. Notice the shape of your hand and try to imagine how little you could do if the framework was one large flat bone instead of nineteen long ones! Now look at a long bone and notice that the central part is narrower than the ends. This is the shaft of the bone and is hollow. The ends are called the extremities and are bulbous (Fig. 2.1).

Short bones
These have no definite shape, but are collected together in groups at the wrists and ankles where they are strong enough to support the weight of the body (Fig. 2.2).

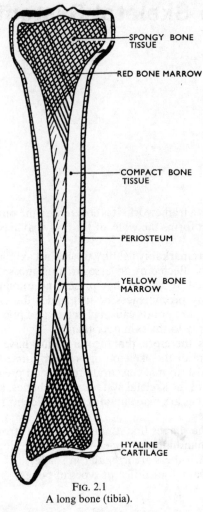

Fig. 2.1
A long bone (tibia).

Irregular bones
These resemble the short bones but have holes in them through which pass important structures such as blood vessels and nerves (Fig. 2.3).

Flat bones
These are found in the skull, the ribs and the pelvis, forming the walls of the cavities of the head and trunk and protecting their contents (Fig. 2.4).

All bones are roughened where muscles are attached, smooth where joints are formed, and grooved for nerves and blood vessels.

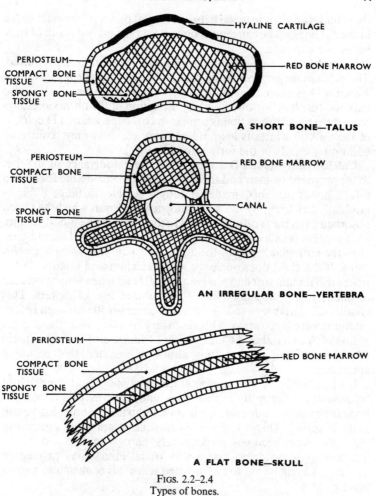

FIGS. 2.2–2.4
Types of bones.

THE TISSUES WHICH FORM THE BONES

These are:
    Bone tissue (compact and spongy or cancellous).
    Fibrous tissue.
    Hyaline cartilage.
    Fatty tissue (Figs. 2.1–2.4).

Bone tissue

There are two types of bone tissue, the hard outer **compact** shell which gives the bone its strength and the **spongy or cancellous** tissue to be found in the extremities of the long bones, in the centre of the short bones and sandwiched between the flat plates of compact tissue in the flat bones. The cancellous bone tissue is hard but is full of little holes like a sponge.

To make bone tissue hard, calcium and vitamin D are required. The calcium we get from milk, cheese, butter, and green vegetables. Vitamin D is present in milk, butter, and cod liver oil. We also manufacture it in our own skins by the action of the ultra-violet rays from the sun. As this action is similar in cows, the vitamin D content of the milk in winter is low and growing children may require an additional supply in the form of cod liver oil.

Calcium and **vitamin D** are of particular importance in the diet of the pregnant woman and the growing child because the foetal and infant bones are soft, made of tough flexible cartilage. This is gradually hardened by the laying down of calcium which has been absorbed from the food and carried to the bones in the blood stream. This process is called **ossification,** and is not completed in all bones until the individual reaches the age of 18 to 20 years, when growth stops. If the child does not get enough calcium and vitamin D the bones will remain soft and pliable and will bend when weight is taken on them giving the knock-knees and bowed legs of rickets. This condition is rarely seen today but it was common 50 years ago before vitamins were discovered. This discovery together with the organization of Antenatal and Child Welfare Clinics, and the subsequent education of the mother, has almost eliminated this crippling condition.

Broken bones are uncommon in the newly born and young child because of the flexibility of the bones, and when they do occur, the bone snaps on one side and then tears lengthwise like breaking young twigs in spring. This is called a green-stick fracture. As one grows older the bone becomes progressively harder, and in old age it becomes quite brittle so that a very trivial injury may produce a fracture. A bone is said to be fractured when it is broken into two or more parts.

Fibrous tissue
Every bone has a covering of fibrous tissue on the outside. This is a protective covering which helps in the nourishment and growth of the bone and it is called the periosteum. It completely surrounds the bone except for the extremities.

Observe the small openings in the bone which look like scratches on the shaft but appear bigger at the ends. This is where the blood vessels penetrate the bone. Bones being living organs, must have a good blood supply.

Hyaline cartilage
This is a smooth shiny material which allows the bones to glide on

each other so forming movable joints. It covers the extremities of the long bones.

Fatty tissue or marrow
The fatty tissue in a bone is found in the spaces in the spongy bone where it is known as red bone marrow. This is where the red blood cells are produced. There is also fatty tissue in the hollow shafts of the long bones; this is yellow bone marrow. It contains very few blood cells and is mainly a fat store.

The Bones of the Skeleton

The skeleton is made up of a number of parts. The bones of the head and trunk form the central support. The mobile parts consist of the upper limbs which are attached to the trunk by the shoulder girdle and the lower limbs which are attached to the trunk by the pelvic girdle. (Fig. 2.5.)

Before going on to study the individual bones the following descriptive terms must be fully understood.

ANATOMICAL TERMS

The *Anatomical Position* is standing upright with the face looking directly forwards, the legs straight with the feet together and the arms by the side with the palms of the hands facing the front.

The *Midline* is an imaginary line drawn through the centre of the body.
Superior means above.
Inferior means below.
Anterior means in front.
Posterior means behind.
Lateral means farthest from the midline.
Medial means nearest to the midline.

THE SPINE OR VERTEBRAL COLUMN

The human animal is a vertebrate. This means he has a backbone. This is not one long bone extending from the head to the tail, but a series of small irregular bones all attached to each other in such a way that the spine can carry out its function as a central support for the body and yet remain very mobile.

These irregular bones are called **vertebrae.** There are thirty-three of them and with a few exceptions they are similar in structure. Each one has a box-shaped body which lies in front and takes the weight. Projecting backwards from this is an arch of bone, called the neural arch because it encloses a canal through which passes the spinal cord

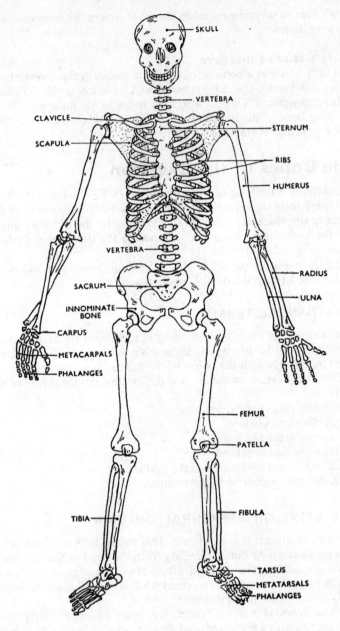

Fig. 2.5
The bones of the skeleton.

which carries the nerves from the brain to all parts of the body. This arch has three processes projecting from it to which are attached the muscles which move the spine, the most pronounced process being the one which projects to the back and lies just under the skin (Fig. 2.6). These are the spinous processes and form important pressure areas when your patient must be nursed on his back in what is called the supine position.

THE DIFFERENT REGIONS AND CURVES OF THE SPINE

If the spine is viewed from the side it will be seen that it is not straight but forms four curves. These are not all present at birth but gradually develop as the infant unfolds, lifts its head up to look around, sits up on its mother's knee, crawls, stands up and eventually takes its

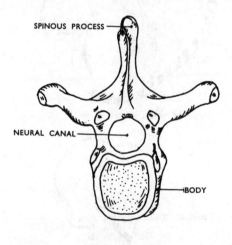

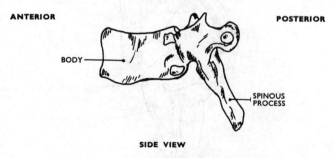

Fig. 2.6
A typical vertebra (dorsal).

first steps. When the baby is born and for the first few weeks of life his spine consists of one continuous curve so that he is curled up like a little ball. As his brain develops so does the desire to see what is going on and he looks upwards. This develops a backwards curve

Fig 2.7   THE SPINAL CURVE AT BIRTH

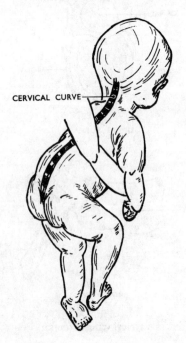

Fig 2.8

in the neck. By the time he has reached the stage of pulling himself up from all-fours to the upright position the fourth curve has developed in the small of his back (Figs. 2.7, 2.8, 2.9).

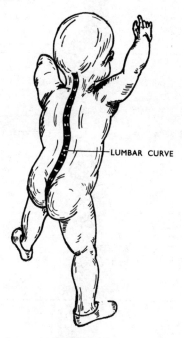

FIG 2.9

FIGS. 2.7–2.9
The curves of the spine.

There are five regions in the spine made up as follows:
The neck—the **cervical** vertebrae.
The back—the **thoracic** vertebrae.
The loin—the **lumbar** vertebrae.
The tail—the **sacral** and **coccygeal** vertebrae.

The bones forming these different parts of the spine are all slightly different from each other and they can be distinguished by their size and shape. The smallest bones are at the top and they become progressively bigger until the pelvis is reached then smaller again to form the tail (Fig. 2.10).

The cervical vertebrae

There are seven of these in the neck. They are smaller than the rest and have three canals through them, the neural canal and two smaller ones which carry blood vessels up to the brain. The spinous processes have two points to them and are said to be forked. The first two cervical vertebrae are called the **atlas** and the **axis**. The atlas is a round ring

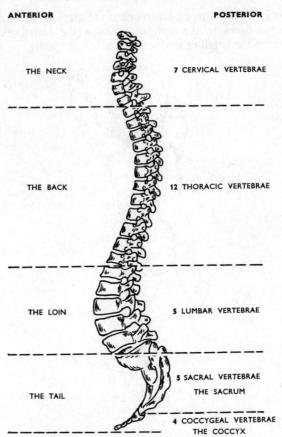

Fig. 2.10
The regions and curves of the spine.

of bone with no body but a large canal for the beginning of the spinal cord. It carries the skull on its shoulders and because of its shape allows nodding movements of the head. The axis has a tooth-like process projecting upwards from its body. This is the pivot round which the atlas and the skull revolve giving the shaking movements of the head.

The dorsal or thoracic vertebrae

There are twelve of these. They form the back of the thorax and have the ribs attached to them. They are slightly larger than the cervical vertebrae because they carry more weight and have very sharp spinous processes. These are very easily felt under the skin.

The lumbar vertebrae
There are five of these. They are the largest as they have to take the weight of the trunk, head, and upper limbs. The spinous processes are much less pointed but are still easily identified in the lower part of the body.

The sacrum
There are five sacral vertebrae fused together to form a triangular-shaped bone called the sacrum. This is wedged between the pelvic bones at the back of the pelvis. There is very little muscle over this bone and when lying in bed this area is subjected to a considerable amount of pressure. It is important, therefore, that all patients who are confined to bed should have the skin over the sacrum and buttocks massaged regularly. This is to ensure that the patient does not get a 'bedsore', a grave complication which retards the patient's recovery and should not occur.

The coccyx
The last four vertebrae are not properly formed. They are also fused together and form the coccyx. It has no function in the human but in the lower animals it is the tail.

HOW THE VERTEBRAE ARE HELD TOGETHER

To understand this we must keep in mind the functions of the spinal column. These are:
  Protection.
  Movement.
  Support.
  The last two seem to contradict each other because we know that the best type of support is something rigid, but a series of ligaments, muscles and discs keep our spines both rigid and movable.

Protection
The delicate spinal cord extends the whole length of the neural canal from the first cervical vertebra to the lumbar region, yet we must bend and twist our spines without harming it.

Holding the vertebrae together and preventing them from slipping off each other are a series of strong ligaments which act as ties and allow the necessary range of movement and no more. In any injury to the spine these ligaments may be torn allowing the vertebrae to move away from each other, thus disrupting the continuous neural canal and damaging the cord. First Aiders and nurses should therefore be very careful when lifting any patient with a history of having hurt his back. Such patients should be kept flat, and if they must be moved they should be rolled or lifted in one piece, never bent in the middle. As the spinal cord carries the nerves from the brain to all parts of the body injury will cause the distressing condition known as paraplegia when there is paralysis of everything below the level of the injury.

Movement

Our spines are very mobile giving a wide range of movement, forwards and backwards, from side to side and twisting. These movements are possible because there are small freely movable joints between the neural arches of all the vertebrae and because the bodies are separated by discs.

These discs act as cushions, rather like the sorbo or air-ring between your patient and the mattress allowing him just enough movement to vary the area of pressure from one part of his buttocks to another. When the spine is bent forwards the area of pressure is at the front of the disc, when bent backwards it is at the back of the disc and at the side when bent sideways. Discs, like cushions, sometimes get worn or torn and the soft inner part leaks out. This causes pressure on the nerves leaving the cord, with the resulting pain of the so-called 'slipped disc'.

Support

There are muscles attached to the spine for the whole of its length. These produce the movements and allow us to fix our spines in whatever position we desire so that it may act as a rigid support for the body. In complete paralysis of these muscles the spine can not be held in the upright position and is therefore no longer a support. Paralysis of the muscles on one side will allow the strong muscles on the other side to pull the spine over, rather like the broken guy rope of a tent.

THE SKULL

This is described in two parts:
(1) The *Cranium,*
(2) The *Face* (Figs. 2.11 and 2.12).

THE CRANIUM

This is the bony box which contains and protects the brain. It consists of eight mainly flat bones jointed together to form the walls of the cranial cavity.

The frontal bone

This forms the forehead and the roof of the orbits—the cavities of the face containing the eyes.

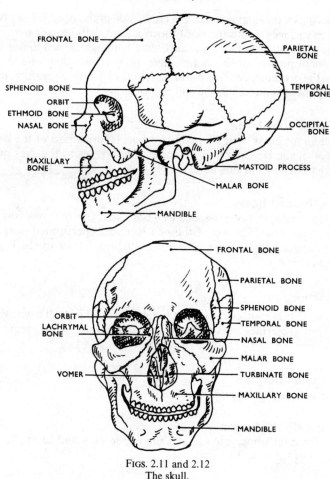

Figs. 2.11 and 2.12
The skull.

The two parietal bones
These form the top of the skull.

The two temporal bones
These contain the delicate internal part of the ears. On the outside of the bone can be seen the opening into the ear behind which lies the prominent mastoid process which is occasionally a site of infection—mastoiditis.

The occipital bone
This forms the back of the cranium and contains a large opening called the foramen magnum where the spinal cord leaves the brain

to pass down the neural canal. On either side of the opening are two smooth surfaces where the skull forms a joint with the atlas.

It is useful to remember that any head bandage carried under the prominent portion of this bone is less likely to slip than one taken above the prominence. This prominence may become a pressure area if the patient is nursed in the supine position.

The sphenoid bone
This is an irregular bone shaped like a bat. It forms part of the base or floor of the cranium and the large optic nerves to the eyes pass through openings in this bone.

The ethmoid bone
This is an irregular bone shaped like a box and forms the bony framework of the nose. The top of the box is a small perforated portion through which the endings of the nerve of smell pass to the brain from the nose.

THE FACE
This consists of fourteen irregular bones arranged in such a way that they form the orbits, part of the nose and the walls of the mouth or oral cavity.

The malar bones
These are the cheek bones.

The maxillary bones
These form the upper jaw, contain the upper teeth and form part of the roof of the mouth.

The mandible
This is the lower jaw, contains the lower teeth and is attached to the temporal bone by two freely movable joints which can be felt just in front of the ear when chewing.

The nasal bones
These form the bridge of the nose.

The lachrymal bones
These are very small and are situated in the inner corner of each orbit. They contain the tear ducts.

The turbinates and the vomer
These form part of the walls and the septum of the nose.

The palatine bones
These form the back of the roof of the mouth or palate.

THE JOINTS OF THE SKULL

The joints of the skull, with the exception of the jaw, are immovable joints or sutures. The edges of the bones are very finely serrated, as with a fine fret saw, and they fit together like the pieces of a jig-saw puzzle.

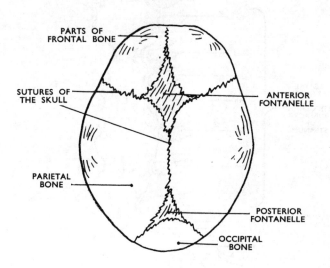

Fig. 2.13
The skull in early infancy. The fontanelles.

Where the sutures of the frontal and parietal bones join in an infant there is a soft area which does not ossify until about the eighteenth month. This is called the anterior fontanelle and is the easiest site to take the pulse of an infant as the blood vessels can be felt pulsating through the soft membrane. There is a similar fontanelle at the back but this ossifies in the first few weeks of life. These soft areas allow the skull to be moulded during birth (Fig. 2.13).

THE SINUSES OF THE SKULL

The sinuses of the skull are spaces in some of the bones. These spaces communicate with the nose, contain air, help with production of voice and keep the skull light.

Sinusitis is infection in these spaces which has spread from the nose or throat. There are sinuses in the frontal, maxillary, ethmoid and sphenoid bones. The frontal sinuses are situated immediately above the orbits and the maxillary sinuses—or antra of Highmore—just below the orbits on each side of the nose.

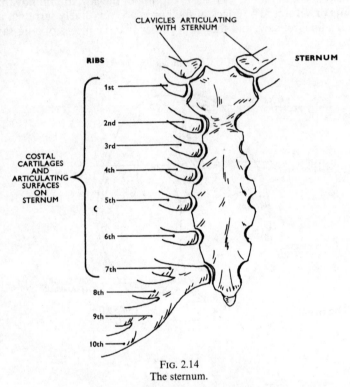

Fig. 2.14
The sternum.

THE THORACIC CAGE OR THORAX

The thoracic cage forms the upper cavity of the trunk which contains and protects the heart and lungs. At the back are the thoracic vertebrae and attached to these are twelve pairs of ribs. The ribs form the sides and front of the cage and are attached by cartilages to the sternum.

The sternum (Fig. 2.14)

The sternum is the breast bone. It lies just under the skin and can be felt easily. It is a flat bone shaped like a dagger.

It is a very light bone, being mainly composed of spongy tissue. It contains an easily accessible supply of red bone marrow. A sternal puncture is the passing of a wide bore needle into the sternum in order to remove some red bone marrow. This is done in certain types of anaemia to examine the state of the red blood corpuscles during their development.

This flat bone protects the heart which lies immediately beneath it. External cardiac massage is a means of stimulating the heart to beat again after it has stopped by rhythmically pressing on the sternum and thus massaging the heart wall until it starts to beat on its own once more.

The ribs (Fig. 2.15)

There are twelve pairs of ribs. They are flat curved bones. At the back they are attached to the thoracic vertebrae, curve backwards then forwards, and with the exception of the last two pairs, are attached to the sternum by pieces of cartilage—the costal cartilages. These cartilages give a spring to the thorax which helps to prevent fractures occurring as a result of a blow on the chest. If the violence is severe enough the ribs will fracture, and it must be remembered that such an injury may be complicated by damage to the lungs.

The anterior and posterior attachments of the ribs are freely movable allowing easy upwards and outwards movements of the chest on breathing.

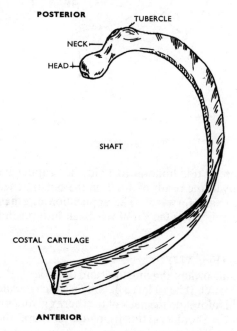

FIG. 2.15
A rib (left).

THE SHOULDER GIRDLE

The shoulder girdle consists of two clavicles or collar bones and two scapulae or shoulder blades forming a girdle round the upper part of the thorax. The girdle is incomplete, there being a gap between the scapulae at the back. The function of the shoulder girdle is to give greater freedom of movement to the upper limbs.

The clavicle (Fig. 2.16)
The clavicle is an 'S' shaped bone which can be felt along the whole of its length if you run your finger from the prominence where it is attached to the sternum, to the shoulder where it has an attachment to the scapula.

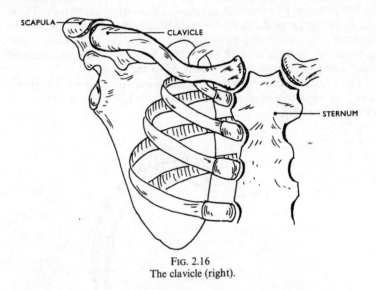

FIG. 2.16
The clavicle (right).

The function of this bone is to hold the scapula back. If it is broken, usually as the result of a fall on the outstretched hand, the whole shoulder falls forwards. The application of a figure-of-eight bandage aims at bracing the shoulders back into position until the bone heals.

The scapula (Fig. 2.17)
The word scapula means 'digging tool' and this bone is shaped rather like a garden trowel. It has a large flat blade which is called the body of the bone. This moves on the back of the thorax giving the shrugging movements of the shoulders. It is triangular in shape and curved to fit over the ribs. The outer border of the bone forms one of the boundaries of the space under the shoulder called the axilla. At the

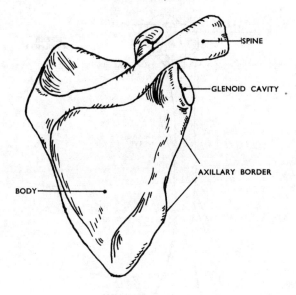

POSTERIOR SURFACE

Fig. 2.17
The scapula (right).

top of this axillary border is a small saucer-shaped surface which is covered with hyaline cartilage and forms the shoulder joint with the humerus. It is called the glenoid cavity.

Projecting from the back of the bone is the spine—the handle of the trowel—which lies just under the skin and, like the clavicle, can be felt along its whole length from the middle of the back to where it unites with the clavicle above the shoulder joint.

The spine of the scapula is an important pressure area especially if the patient is very thin and must be nursed on his back in the supine position.

THE UPPER LIMB

This consists of one bone in the arm, the humerus; two bones in the forearm, the radius and the ulna; the carpal bones of the wrist; and the metacarpals and phalanges, the bones of the hands and the fingers.

The humerus (Fig. 2.18)

The humerus is a long bone with a shaft and two extremities. The upper extremity has a rounded head which forms with the glenoid cavity of the scapula, the shoulder joint. Lateral to this is a rough

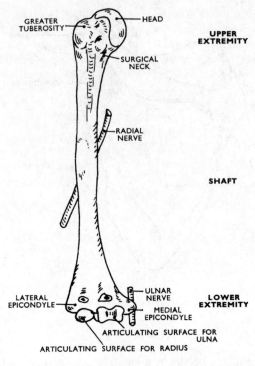

Fig. 2.18
The humerus (right).

process called the greater tuberosity. This, although fairly well covered with muscle, can be felt just under the tip of the shoulder and forms an important pressure area when the patient is nursed in the true lateral position.

Below the upper extremity the bone narrows to form the shaft. The narrowest part is called the surgical neck as in injuries to the shoulder this is the part which is commonly broken. An important nerve, the radial nerve, winds round the back of the shaft of the humerus. This nerve carries messages, or impulses, to the muscles which pull up the wrist and straighten the fingers. If it is injured the messages can no longer pass along the nerve. Compare this with a telephone when the wires are cut and the telephone goes dead. In the same way the muscles become paralysed when the messages which make them move can no longer reach them. The result is a dropped wrist, a most incapacitating deformity. Careless handling of the unconscious patient may produce this deformity. When placing such a patient on a theatre table or trolley the nurse must make sure that the arms are well supported and not allowed to hang over the edge. If this does happen and the arm is allowed to remain in this position, the edge of the table or trolley will press on the radial nerve and injure it.

The lower extremity of the humerus is broader and flatter than the upper extremity. It has two smooth surfaces and two rough projections, or epicondyles. The smooth surfaces join with the radius and ulna to form the elbow joint. The epicondyles project out on either side and if the patient is bedridden they require regular attention as they become sore from friction with the bedclothes. Another important nerve winds round the back of the inner epicondyle and if you press your fingers in between this and the point of the elbow, this nerve—the ulnar nerve—can be felt like a wire, and can be rolled between the fingers. Pressure or a knock on this point causes tingling in the ring and little fingers, hence the term 'funny bone'.

THE BONES OF THE FOREARM
The radius (Fig. 2.19)
The radius is a long bone, the outer bone of the forearm. With the

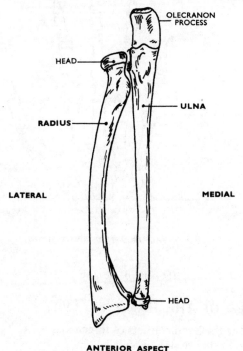

ANTERIOR ASPECT
FIG. 2.19
The radius and ulna (right).

hand lying palm upwards it is the bone in line with the thumb. The upper extremity is a rounded head shaped like a button. This joins with the humerus to form part of the elbow joint and also with the ulna to give the important twisting movements of the forearm.

The lower extremity is larger and flatter than the upper extremity and the radial artery passes over the front of it into the hand. At this point the artery can be easily compressed against the bone and therefore it is the most convenient site to count the pulse.

The ulna (Fig. 2.19)

The ulna is a long bone, the inner bone of the forearm. It has a claw-shaped upper extremity which forms the main part of the elbow joint with the humerus. The upper part of the claw is the olecranon process and forms the point of the elbow. This is another site where the skin very quickly becomes red from friction with the sheets.

The lower extremity is a small rounded head which can be seen at the back of the wrist in line with the little finger.

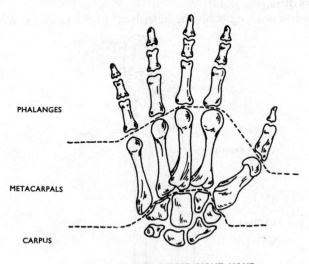

FIG. 2.20
The bones of the hand.

THE BONES OF THE HAND (Fig. 2.20)

The skeleton of the hand consists of three parts:
  The *Carpus* at the wrist;
  The *Metacarpals* in the palm;
  The *Phalanges* in the fingers.

The carpus
The carpus consists of eight short bones, the carpal bones, arranged roughly in two rows of four. This arrangement gives strength to the wrists which can, when necessary, support the whole weight of the body.

The metacarpals
The metacarpals are long bones. There are five of them and their rounded heads form the knuckles.

The phalanges
The phalanges are very small long bones, there being three in each finger and two in the thumb.

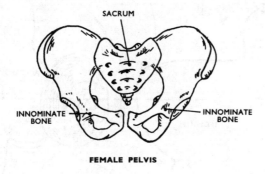

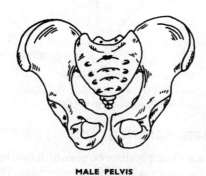

Fig. 2.21
The pelvic girdle.

THE PELVIC GIRDLE

The pelvic girdle consists of two irregular bones, the innominate bones. These bones join the sacrum at the back to form the pelvic cavity or pelvis (Fig. 2.21). As this cavity contains the organs of reproduction the formation of the female pelvis differs from that of the male. The male pelvis is much smaller, narrower and more funnel shaped than the female pelvis which is large and wide to allow the birth of a child at the end of pregnancy.

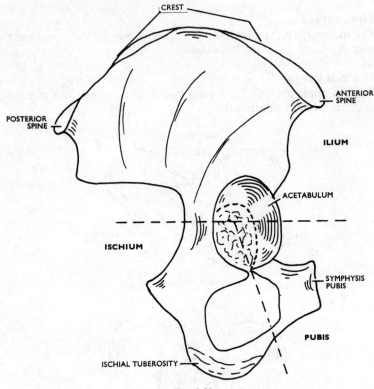

Fig. 2.22
The innominate bone (right).

THE INNOMINATE BONE (Fig. 2.22)

The innominate bone is really three bones which fuse together during childhood. The upper part is flat, the rest irregular. The three bones are called:
  Ilium,
  Ischium,
  Pubis.

The ilium
This is the flat upper part which has the powerful hip muscles of the buttocks attached to it. At the top of the bone is a long curved crest

which ends in sharp spines at the back and front. The large notch at the back of the ilium is where the **sciatic nerve** comes out of the pelvis into the buttock. This area of the bone is called the greater sciatic notch.

The posterior spine forms a pressure area with the sacrum when the patient is nursed in the recumbent position. In the less common prone or face-down position the skin over the anterior spine requires a great deal of attention. These pressure areas require particular care when you are nursing a paralysed patient, or any patient on a plaster bed. If the patient is paralysed and nursed prone in bed, the sacral curve must be supported but the pillows must be placed in such a way that the anterior iliac spine and crest are free from undue pressure. Always remember that a paralysed patient may not be able to feel and therefore will not complain of pain or discomfort in the region of any bony prominence (Figs. 11.9 and 11.12).

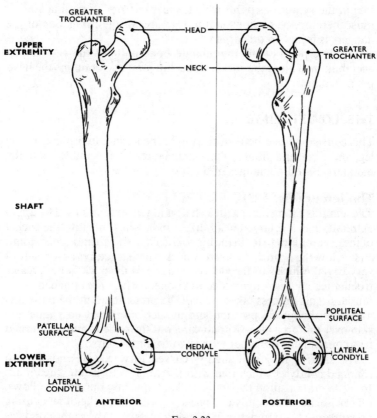

Fig. 2.23
The femur (right).

The ischium
This is the lower posterior portion of the bone. It has a rough tuberosity which takes the weight of the body in the sitting position. These tuberosities are protected by small sacs of fluid, called bursae, which act as water cushions. It is therefore possible to maintain the sitting position for a considerable period of time without friction occurring between bone and skin. It must be remembered, however, that these points require care and attention when the patient is thin and confined to a wheel chair. Such patients, if their arms are strong enough, must be taught to raise themselves up once every five to ten minutes to change their position slightly so that the same area of skin is not continuously receiving pressure and therefore being deprived of its blood supply.

The pubis
This is the thin front portion of the bone. The two innominate bones join at the symphysis pubis which is just in front of the bladder. It must therefore be obvious that in fractures of the pelvis one of the dangers is injury to this organ.

All three parts of the innominate bone unite at the cup-shaped acetabulum. This is the socket of the hip joint and contains the head of the femur.

THE LOWER LIMB

This consists of one bone in the thigh, the femur; two bones in the leg, the tibia and fibula; the tarsal bones of the ankle, and the metatarsals and phalanges of the foot and toes.

The femur (Fig. 2.23)
The femur is a long bone with a shaft and two extremities. The upper extremity has a rounded head, like a ball, which fits into the socket of the acetabulum to form the hip joint. This is a ball and socket joint. Just below the head is the narrow neck which fractures as a result of very trivial injuries to the elderly. Lateral to the neck is the greater trochanter, a large rough process of bone to which are attached many muscles. This process takes most of the pressure when the patient is nursed in the lateral position and in such cases, not only must the skin over this area receive great care, but the patient must be turned from one side to the other every two to three hours.

The shaft is long and slightly curved forwards. It ends in two rounded condyles which join with the tibia to form the knee joint. In the lateral position they must be kept apart by the use of pillows, otherwise the inner condyle of one knee will rub on the other causing pressure and friction between two skin surfaces. Above the condyles in front is the smooth patellar surface and at the back the triangular

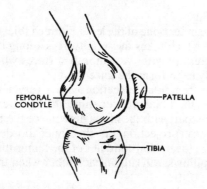

THE PATELLA

Fig. 2.24
The patella (right).

shaped popliteal surface. The popliteal space is the area behind the knee through which blood vessels and nerves pass into the leg.

Pressure on a blood vessel slows the circulation of blood and tends to make it clot. The resulting thrombosis is a serious post-operative complication which is encouraged by the placing of a large firm pillow under the knees. If possible every patient must be encouraged to move around in bed and exercise the legs. If the upright position must be maintained this should be done by using a foot rest and not a knee pillow (Fig. 11.15).

The patella (Fig. 2.24)

The patella is a sesamoid bone, that is, a bone developed in the tendon of a muscle. It is triangular in shape and forms the knee cap. It takes the weight in the kneeling position. Like the ischial tuberosity it is protected from friction by a small bursa.

The patellae form pressure areas in either the prone or supine position. In the former they must be protected from friction with the bed, and in the latter, especially if flexion deformities of the knees are present, a bedcage must be used to support the weight of the bedclothes.

The tibia (Fig. 2.25)

The tibia is the inner bone of the leg. It is much thicker than the fibula and is the one which takes the weight. It is a long bone with a thick flat upper extremity with two smooth surfaces which join with the femoral condyles to form the knee joint.

The shaft is triangular in section with a very superficial anterior border. This is the tibial crest or shin. The lower extremity helps to form the ankle joint with the fibula and one of the tarsal bones. The prominent bony projection on the inner border is the medial malleolus. Like the medial condyles of the femur the malleoli, unless separated by pillows, will rub on each other when the patient is lying in the lateral position.

The fibula (Fig. 2.25)

The fibula is the outer bone of the leg. The upper extremity of this long bone is a rounded head which can be felt on the outside of the knee and slightly to the back. The important lateral popliteal nerve winds round this prominence. It is easily damaged by pressure and, as it supplies the muscles which pull up the foot, injury will cause drop foot similar to the deformity resulting from too tight bedclothes.

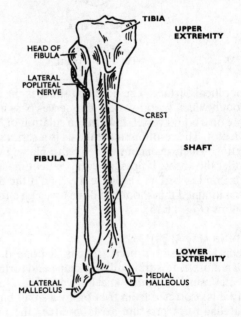

ANTERIOR ASPECT
Fig. 2.25
Tibia and fibula (right).

This condition is very disabling and must not be allowed to occur. When applying a knee bandage care must be taken to see that it is not too tight; that it is fastened at the front and not at the side and that knots are avoided. A hard pad or pillow must never be placed under the knee and this area should always be protected from the pressure of leg splints.

The lower extremity of the fibula is the lateral malleolus. It is the outer bone of the ankle joint and like the medial malleolus may be subject to pressure and friction.

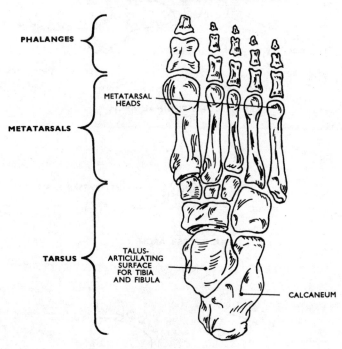

FIG. 2.26
The bones of the foot.

THE BONES OF THE FOOT (Fig. 2.26)

The skeleton of the foot consists of three parts:
 The *Tarsus* or bones of the ankle and heel.
 The *Metatarsals* or bones of the foot.
 The *Phalanges* or bones of the toes.

The tarsus
The tarsus consists of seven short bones, the tarsal bones, which take

the weight of the body in the standing position. Two of these bones are larger than the others, the **talus,** the highest, is the third bone involved in forming the ankle joint. The **calcaneum,** the largest bone, forms the heel.

The metatarsals

The five metatarsals are long bones. Their heads form the ball of the foot.

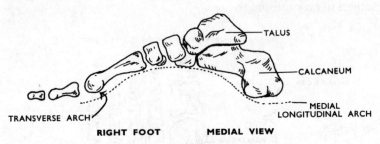

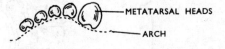

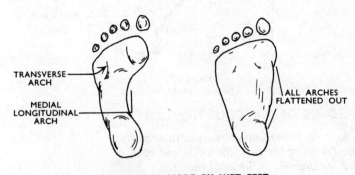

Fig. 2.27
The arches of the foot.

The phalanges

These are fourteen very small long bones forming the toes. As in the thumb, the big toe has only two phalanges whilst the other toes have three.

THE ARCHES OF THE FOOT (Fig. 2.27)

The bones of the foot are arranged in such a way that they form three arches, two lengthwise and one across the foot. These arches give the foot its spring, and if they fall the resulting gait will be ungainly and the feet painful. There are many causes of flat feet, amongst them, the habitual wearing of badly made slip-on shoes. Patients who have been a long time in bed and young people who have just left school to take some job which requires long periods of standing and walking, require good supporting shoes, otherwise they may develop painful fallen arches.

THE PRESSURE AREAS OF THE FOOT

These are the toes and the heels. The bedclothes press on the toes and the heels rub against the sheet. A bedcage will take the weight of the bedclothes off the toes and pressure can be removed from the heels by the use of small pillows.

Questions

(1) One of the following tissues is not part of the structure of a bone. Which one? (a) cartilage (b) epithelium (c) adipose tissue (d) fibrous tissue.
(2) .... bone tissue forms the hard outer shell of the bone.
(3) The spaces in the cancellous bone tissue contain ....
     ....
(4) Yellow bone marrow is found in: (a) long (b) short (c) flat (d) irregular bones.
(5) In rickets the bones are soft because there is not enough .... and vitamin .... in the diet.
(6) If a child is to have strong bones there must be plenty of (a) iron (b) meat (c) milk (d) salt—in the diet.
(7) In the anatomical position the radius is .... to the ulna.
(8) The abdominal cavity is superior to the .... cavity.
(9) Anterior is the word which means ....
(10) Vertebrae are: (a) long (b) short (c) irregular—bones.
(11) The number of vertebrae in the spine are: (a) 30 (b) 33 (c) 32 (d) 25.

Anatomy and Physiology as Applied to Nursing

(12) The most prominent part of a vertebra is its ....
....
(13) The part of the vertebra which takes weight is the ....
(14) Which of the following regions of the spine has the most prominent spinous process: (a) cervical (b) thoracic (c) lumbar?
(15) The sacral curve of the spine is developed ....
....
(16) There are .... cervical vertebrae. They form the skeleton of the ....
(17) In shaking movements of the head the ....
revolves round the ....
(18) The ribs are attached to the: (a) cervical (b) lumbar (c) dorsal (d) sacral—vertebrae.
(19) The functions of the spinal column are: (a) ....
.... (b) ....(c) ....
(20) The vertebral column protects the ....
(21) The bodies of the vertebrae are separated by ....
....
(22) Most of the bones of the cranium are .... bones.
(23) The bone at the back of the cranium is the..........bone.
(24) The optic nerve passes through: (a) the ethmoid bone (b) the temporal bones (c) the lachrymal bones (d) the sphenoid bone.
(25) The ear is situated in the: (a) occipital (b) temporal (c) patietal—bone.
(26) The upper jaw is formed by the .... bones.
(27) The only movable joint in the skull is the joint between the mandible and the .... All the other joints are called ....
(28) Which of the following bones contain sinuses: (a) frontal (b) parietal (c) occipital (d) lachrymal?
(29) The sternum articulates with two .... and seven pairs of ....
(30) The glenoid cavity is part of the ....
(31) Pressure applied to the back of the arm may damage the .... nerve. This will result in a deformity called ....
(32) The olecranon process is the upper end of the ....
(33) There are .... phalanges in the fingers. These are .... bones.
(34) The part of the pelvis which takes the weight when sitting is the ....
(35) The femur articulates with the .... of the innominate bone.
(36) The bladder lies behind: (a) the iliac spine (b) the symphysis pubis (c) the ischial tuberosity.

The Skeletal System

(37) The prominent part of the femur in the region of the hip joint is the ....
(38) The patella is a: (a) long (b) short (c) sesamoid (d) irregular—bone.
(39) The tibia is the: (a) medial (b) lateral (c) anterior—bone of the leg.
(40) One of the following is a cause of drop foot. Which one? (a) bad footwear (b) pressure on the lateral side of the knee (c) pressure on the medial side of the knee.
(41) The lateral malleolus is part of the ....
(42) There are .... tarsal bones, the largest one forms the heel and is called the ....

3. The Joints and Muscles

Having studied the framework of the body, we must now try to find out how it is joined together and what makes it move. The bones, the joints, the muscles and the nerves form the **locomotor system**.

The Joints or Articulations

There are three types of joints holding the bones together:
> Immovable.
> Slightly movable.
> Freely movable.

Immovable joints can be seen in the sutures of the skull where the bones are soundly fixed together by fibrous tissue (Fig. 3.1).

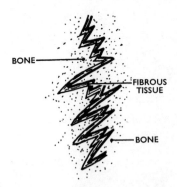

FIG. 3.1
Fixed or immovable joint. (Sutures of the skull).

Slightly movable joints are to be found between the bodies of the vertebrae where a pad of cartilage allows slight movement (Fig. 3.2).

Most of the other joints of the body are **freely movable.** They are sometimes called **synovial joints** because of the sticky fluid, like the

The Joints and Muscles   43

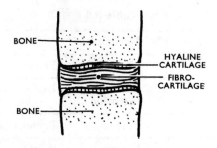

Fig. 3.2
Slightly movable joint. (Between vertebrae.)

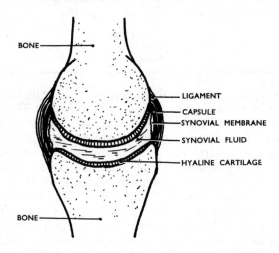

Fig. 3.3
Freely movable or synovial joint.

white of an egg (syn—like, ovum—egg), which lubricates them. These joints are capable of various degrees of movement (Fig. 3.3).

The terms used to describe the movements are:
Gliding—slipping.
Flexion—bending.
Extension—straightening.
Rotation—turning.
Adduction—moving towards the midline.
Abduction—moving away from the midline.

The larger and more important synovial joints of the body are classified according to their movements, as follows (Table 3.1).

Table 3.1

Class	Movements	Joints
Gliding	Slipping	Spine
Hinge	Flexion Extension	Elbow; ankle; knee
Pivot	Rotation	Radio-ulnar
Condyloid	Flexion Extension Abduction Adduction	Wrist Jaw
Ball and Socket	Flexion Extension Abduction Adduction Rotation	Hip Shoulder

If you examine any one of these joints you will see that the surfaces covered with hyaline cartilage fit together accurately. There is, however, one exception, the knee joint where the condyles of the femur are round and the upper extremity of the tibia flat. Little half-moon-shaped wedges of cartilage lie between the two surfaces, shaping the upper end of the tibia and rounding the surfaces to fit the femoral condyles. These cartilages may be torn as a result of injury to the knee, a condition commonly found in footballers and miners.

In all synovial joints the bones are held together by a cuff of fibrous tissue. This is called the **capsule,** and in certain situations it is strengthened by ligaments. A sprain is the condition resulting from the tearing of some of the fibres of a ligament. A more serious condition results when the whole ligament is torn, as this may produce a dislocation which is the displacement of the joint surfaces.

The joint capsule has a smooth lining of synovial membrane. This membrane secretes the fluid which keeps the joint lubricated; a type of self lubricating mechanism (Fig. 3.3). It also covers all surfaces inside the joint capsule which are not already covered with hyaline cartilage. This provides continuous smooth surfaces allowing movement without friction. These joints are meant to be moved and fixation for any reason will result in stiffness and pain.

Joints may swell and stiffen for various reasons and this condition is called **arthritis.** During the acute stage of arthritis pain is very severe and the patient will resist movement. He will lie in the most comfortable position, which is usually on his side, all curled up, with his hips and knees flexed. Unless his joints are moved several times a day the joint capsule will tighten and contractures will be formed.

Such deformities may mean that the patient becomes bedridden and any alteration in position is difficult. Under such circumstances gross bedsores may develop and it is not unknown for an arthritic patient to be admitted to hospital with his knees nearly up to his chin, heels touching his buttocks and feet dropped. To prevent this one must understand the importance of keeping the painful joints mobile (Fig. 3.4).

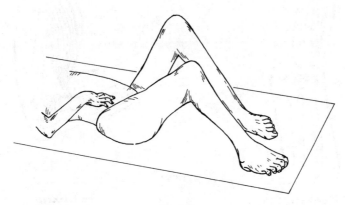

Fig. 3.4
Flexion contractures in a patient with arthritis.

The Muscular System

Without active muscles no voluntary movement can take place at a joint. Because of this the type of tissue forming the muscles attached to the skeleton is called voluntary muscle tissue, *i.e.* it is under the control of the will.

The muscles form the flesh of our bodies and if you take a piece of lean meat and tease it out you will see that it consists of tiny bundles of short thread-like structures. These are the cells of the tissue and they are called muscle fibres. Each one is capable of shortening, or contracting, in order to produce movement. A muscle consists of thousands of these bundles wrapped in a sheet of fibrous tissue and attached to the skeleton by tendons. Each muscle passes over one or more joints and when it contracts it pulls on the bone producing movement (Fig. 3.5). The type of movement depends on the position of the muscle; thus some are extensors and some flexors, some adductors and some abductors.

A muscle must possess two things before it can contract. These are:

An adequate blood supply.
A nerve supply.

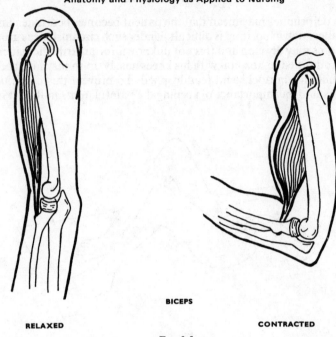

BICEPS

RELAXED   CONTRACTED

Fig. 3.5
Muscle action.

The **blood** brings along glucose and oxygen to the muscle. The glucose comes from the food we eat and the oxygen from the air breathed in by the lungs. The **nerve** supplies the spark, or impulse, which starts the combustion of the glucose with the oxygen. This provides the power or the energy which makes the muscle contract.

In a petrol-driven motor the sparking plug ignites the petrol which burns with the oxygen from the air. This releases the energy which makes the machinery work and the wheels move. As a result of this burning the engine becomes hot and there must be some means of keeping it cool. As long as the engine is running poisonous gases are being produced. These fumes are generally known as the exhaust.

Now compare your own body with this motor. We do not fill ourselves up with petrol, but we do eat food. When we stoke up with such foods as potatoes, bread and sweets, we are providing the muscles with fuel in the form of **glucose.** This glucose can be stored just as the petrol is stored in the tank of the car. When movement is required, **oxygen** which has been brought from the lungs burns with the glucose releasing **energy** and **heat.** The energy makes the muscle contract and so produce movement, and the heat helps to keep our body temperature at 37 deg. C. There is often too much heat produced for our body requirements, therefore we sweat. Sweat is mainly water and this is evaporated from our skins. This evaporation uses up the extra heat but it also uses up water, so we must drink to replace it. What about the exhaust? Two waste gases are formed in our muscles, one is **carbon dioxide** and the other **water vapour.** Both are conveyed by the blood stream to the lungs and breathed out.

During exercise all these processes are intensified because the machine which is our body has more work to do. We breathe in more deeply to get more oxygen, and breathe out as much air as we can, to get rid of the carbon dioxide. We will get hungry and thirsty

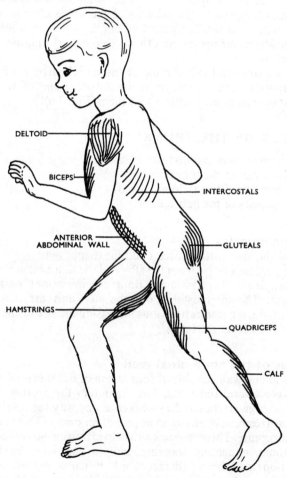

Fig. 3.6
Superficial muscle groups.

because we are using up food and water. It is a well-known fact that exercise is a good appetizer.

Just as a motor car will refuse to move if not supplied with petrol, or if the battery requires recharging, so will our muscles become inactive if not supplied with glucose, oxygen and a nerve impulse. Damage to or blockage of a blood vessel will interfere with the supply of blood carrying the glucose and the oxygen. A tight bandage, particularly a plaster one, is enough to do this and, if left unattended, the muscle will die. Damage or disease of the brain, spinal cord or nerve, will prevent the controlled passage of the impulse and the muscle will be **paralysed.** It is paralysed in contraction if the injury or disease is in the brain or spinal cord, and this is what happens to the muscles of the limbs after a cerebral haemorrhage or stroke. It is paralysed in relaxation when the injury or disease affects the nerves, and this is seen in nerve injuries and in poliomyelitis.

There are some individual muscles and muscle groups which are more important than others from the nursing point of view, and every nurse must know a little about them (Fig. 3.6).

MUSCLES OF THE TRUNK

There are three main groups:
   The muscles of the thorax.
   The muscles of the abdominal wall.
   The muscles of the pelvic floor.

Thoracic muscles
These are the intercostal muscles and the diaphragm.
   The intercostals lie between the ribs and the diaphragm separates the thorax from the abdomen. Both are important muscles of respiration. The intercostals raise the ribs and the diaphragm flattens out during inspiration, thus increasing the size of the thorax so that air can enter the lungs.

Muscles of the abdominal wall
The abdominal wall consists of four layers of flat sheets of muscles which have several functions. They maintain the positions of the abdominal organs; should they weaken in any way the result would be the displacement of an organ or part of an organ. This is called a hernia or rupture. These muscles also help in movements of the spine, in breathing, coughing, sneezing, and in emptying the bladder (micturition) and bowel (defaecation). When a patient is being prepared for abdominal examination his bladder should be empty and he should lie flat on his back with only one pillow. This assures

complete relaxation of the abdominal muscles and allows the doctor to feel and palpate the organs.

Coughing is caused by irritation of the respiratory passages. The abdominal muscles contract, force the abdominal organs against the diaphragm and this pushes more air out of the lungs in an effort to expel the irritating material. Patients who have had abdominal operations try to avoid coughing because any sudden muscular contraction results in a spasm of severe pain.

Muscles of the pelvic floor

These muscles keep the pelvic organs in place. If they weaken a prolapse of the organs may occur. There are three openings

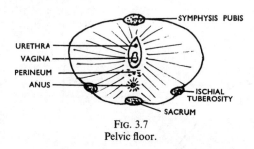

FIG. 3.7
Pelvic floor.

through the female pelvic floor, the urethra leading from the bladder, in front; the vagina from the uterus, in the middle; the anus from the rectum, at the back (Fig. 3.7). The male pelvic floor has only two openings, the urethra in front and the anus behind. The tendinous area between the anus and the anterior openings is the perineum.

OTHER IMPORTANT MUSCLES

The deltoid

This muscle lies over the shoulder joint which it moves. It is sometimes used as a site for intramuscular injections. As it is inserted into the humerus, just where the radial nerve winds round the bone, the nurse must make sure that the injection is given well up towards the shoulder into the bulk of the muscle, where there is no danger of hitting the radial nerve. It is always advisable to remove the patient's arm from his sleeve to ensure the exposure of the whole muscle. As already stated, injury to this nerve will result in a dropped wrist.

The biceps

This muscle lies in front of the upper arm and flexes the elbow. It has a long thick tendon which passes over the elbow joint just to the lateral side of the brachial artery. When blood pressure is to be

The gluteals

These are the muscles of the buttocks. They move the hip joint and play an important part in maintaining the upright position. As these muscles are bulky, they are also used as a site for injections. The large **sciatic nerve,** however, passes through the gluteal muscles on its way down into the leg. It lies towards the inside and in the lower part of the buttock. Therefore the injection must be given into the upper and outer quadrant (Fig. 3.8). This is very important as an injury to the sciatic nerve will produce almost complete paralysis of the lower limb.

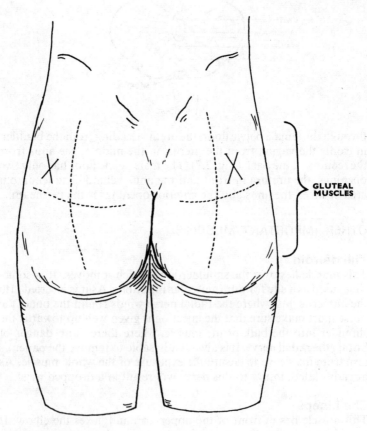

FIG. 3.8
Site for intramuscular injection into the buttock.

The quadriceps

These are the muscles in front of the thigh. They are important in maintaining the upright position and also in walking as they straighten the knee. They should be kept exercised by all bed patients but particularly by those suffering from some of the more common affections of the knee joint, *e.g.* a torn cartilage or rheumatoid arthritis. By maintaining the strength of this muscle the patient, on getting up, will be steadier on his legs.

The outer aspect of the thigh is perhaps the safest part of the whole body for intramuscular injections as there are no large nerves or blood vessels passing through this region.

Questions

(1) The three different types of articulations are: (a) .... (b) .... (c) ....
(2) The sutures of the skull are .... joints.
(3) The intervertebral discs are pads of ....
(4) A freely movable joint is called a .... joint.
(5) Flexion is the same as: (a) bending (b) turning (c) moving towards (d) straightening.
(6) Adduction means ....
(7) The capsule is part of: (a) a fixed joint (b) a slightly movable joint (c) a freely movable joint.
(8) The elbow joint is a synovial .... joint.
(9) One of the following movements occurs at the knee joint: (a) adduction (b) extension (c) abduction.
(10) The wrist joint gives the movements of ....
(11) Skeletal muscles are made of: (a) involuntary (b) voluntary (c) cardiac muscle tissue.
(12) When a muscle shortens it is said to ....
(13) The glucose required for muscle contraction comes from .... foods.
(14) The energy necessary for muscle contraction is derived from the combusion of glucose and ....
(15) One waste product of muscle contraction is: (a) water (b) carbon monoxide (c) glucose (d) oxygen.
(16) Paralysis is: (a) a disease of muscle (b) muscle deprived of its nerve supply (c) muscle deprived of its blood supply.
(17) The .... muscles lie between the ribs and are muscles of ....
(18) The abdominal muscles have many uses one of which is ....
(19) In order to relax the abdominal muscles for examination of

the abdomen the patient should be: (a) upright (b) recumbent (c) prone (d) semi-recumbent.
(20) The anterior opening in the pelvis floor is: (a) the vagina (b) the anus (c) the urethra.
(21) When giving an injection into the deltoid care must be taken because of the position of the .... nerve.
(22) The muscles forming the buttocks are the: (a) gluteals (b) biceps (c) quadriceps (d) deltoid.
(23) The nerve which passes through the buttock is the: (a) popliteal (b) sciatic (c) radial (d) ulnar—nerve.
(24) The muscles in front of the thigh are the .... muscles.

4. The Circulatory System

The **circulatory system** is the means by which food and oxygen are conveyed from the digestive tract and lungs to the body cells. In addition, the circulatory system removes the waste products from the cells to the kidneys, lungs, and skin where they are excreted. It resembles the water supply and drainage system of a community, the difference being that this water supply has the groceries dissolved in it, and it cannot be turned off and on at the tap but must flow continually throughout life!

The organs of the circulatory system are:
　The heart which is the pump.
　The blood vessels which are the pipes.
　The blood which is the fluid carrying the food and waste.

THE HEART

This cone-shaped organ lies in the thorax between the lungs, slightly more towards the left side than the right. It is hollow and made of a special type of muscle called cardiac or heart muscle. Part of this muscle forms a wall or septum dividing the heart into two sides—a right and a left side.

The heart is covered on the outside by a double sac of smooth tissue which has thin lubricating fluid called serous fluid between the surfaces. On the inside is a very smooth epithelium which not only lines the heart but forms two valves dividing the organ into four chambers—two upper and two lower (Fig. 4.1).

All these structures have their own names and functions.

The four chambers

These are the right and left **atria** above, and the right and left **ventricles** below. The blood flows from the atria into the ventricles and is then pumped round the body.

The muscular wall

This is called the **myocardium**. It is made of cardiac muscle tissue. A special type of muscle which unlike the muscles attached to the skeleton, can contract without any control from the brain. It is thicker in the ventricles than in the atria and thickest in the left ventricle. Its function is to contract and squeeze the blood out into the blood vessels then to relax so that the heart can fill up with blood. In the condition called myocarditis, where it is inflamed, the pump is less efficient—it pumps more quickly but with less force. This change will be detected by the nurse when she feels the pulse.

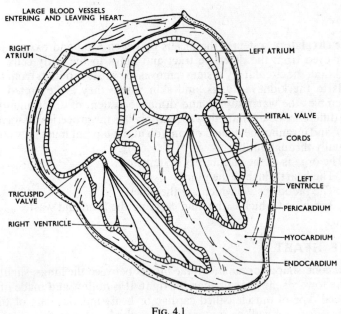

FIG. 4.1
The heart structure.

The outer covering

The outer covering of the heart is called the **pericardium**. It is double and between the two folds is a thin lubricating liquid called serum. Thus when the heart contracts and relaxes it does not rub on the neighbouring lungs. One layer of the pericardium merely glides on the other. In pericarditis the inflamed tissues can be heard rubbing on each other if the sounds of the heart are listened to with a stethoscope.

The inner lining

The inner lining or **endocardium** is smooth to allow the blood to flow over it. The valves prevent the back-flow of blood from the ventricles into the atria. These valves are folds of the endocardium held down by fine cords. There are three folds on the right side and this valve is called the **tricuspid** valve. On the left side there are only two folds so here we have the **bicuspid** valve. This is sometimes called the **mitral** valve because it was thought to resemble a bishop's headdress. These also may be affected by disease. This is valvular disease of the heart—endocarditis—when the valves leak or the opening becomes so small that insufficient blood gets through. This results in a slowing down of the circulation and a collecting of fluid in the tissues and in the lungs. This accounts for the swelling of the ankles (oedema) and the breathlessness seen in these patients.

THE BLOOD VESSELS

There are three types of blood vessels:
    Arteries.
    Capillaries.
    Veins.

The arteries
The arteries leave the heart at the left and right ventricles where they are very large. They branch again and again, getting smaller and spreading out to every part of the body. They are now called **arterioles.** Eventually they become so small and so thin that they cannot be seen except under a microscope. They have now become capillaries.

The capillaries
These form a network throughout all the tissues. They, in turn, join up with each other and become larger, until eventually veins are formed (Fig. 4.2).

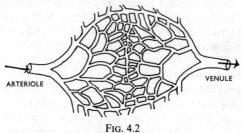

FIG. 4.2
Capillary network.

The veins
The small veins, called **venules,** join other small veins until large veins are formed. These enter the heart at the right and left atria.

The arteries and veins are constructed in a similar way to the heart. They are muscular tubes with a covering and a lining. The lining is smooth, similar to the endocardium, and forms valves in the veins (Fig. 4.3A). The covering is a single sheet of very elastic fibrous tissue, thick in the large arteries and thin in the veins (Fig. 4.3B). The capillaries are very thin and semi-permeable. It is only through the walls of these vessels that nutrients and gases can reach the cells.

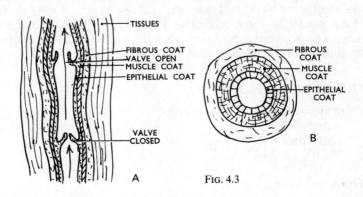

Fig. 4.3

SOME IMPORTANT BLOOD VESSELS

The larger arteries and veins are named and we must know where some of them are before we can take the patient's pulse, or can understand what to do in emergencies when the patient is bleeding.

The coronary circulation

The **aorta** (Fig. 4.4) leaves the heart carrying oxygenated blood round the body. This artery branches again and again until each organ has its own smaller vessel. The branches which supply the walls of the heart are called the **coronary** arteries. These may be blocked by a clot or embolism, a condition referred to as a coronary thrombosis and a common cause of sudden death.

The general circulation

Two **sub-clavian** arteries branch off on each side of the neck and pass into the **axillae** then down the arms where they are renamed the **brachial** arteries. In first aid when there is bleeding from a wound in the hand or fore-arm this artery can be compressed against the humerus in order to arrest the bleeding. The brachial artery divides at the elbow to form the **radial** and **ulnar** arteries. The radial artery can be felt pulsating at the base of the thumb and can tell us quite a lot about the heart beat.

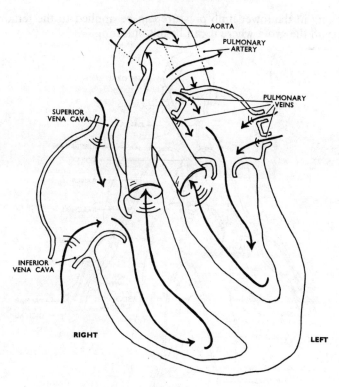

Fig. 4.4
The blood vessels of the heart.

The blood returns from the hand by the veins lying alongside the arteries and sharing their names. The subclavian veins in the neck join to form the **superior vena cava** which enters the right atrium (Fig. 4.4).

The **carotid** arteries pass up the neck to supply the head and brain and the **temporal** and **facial** arteries are branches of these. The temporal can be felt pulsating in front of the ear and the facial, just in front of the angle of the jaw. The large vessels which return blood from the brain are called the **jugular** veins.

All the organs of the trunk have their own arteries which branch from the aorta and their own veins which return the impure blood to the heart by the **inferior vena cava.** The aorta branches in the lower part of the abdomen and two large arteries, the **iliac** arteries, pass out of the trunk into the thigh. These become the **femoral** arteries which travel down the thighs to the back of the knees. They are then called the **popliteal** arteries. The branches of these are the **anterior** and **posterior tibial** arteries which supply the legs and feet. When there is bleeding in the lower limb pressure can be applied to the femoral artery in the groin where it can be felt pulsating.

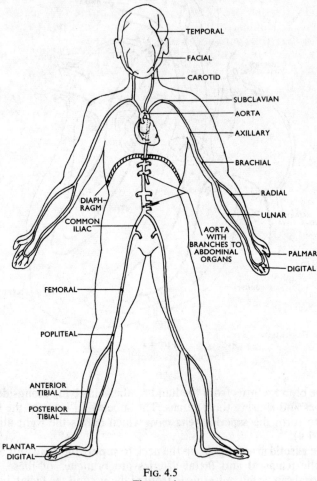

FIG. 4.5
The arteries.

The veins from the lower limbs lie alongside the arteries and unite in the pelvis to form the inferior vena cava which lies to the right of the aorta on the posterior abdominal wall (Figs. 4.5 and 4.7).

The limbs have two sets of veins, the deep ones accompany the arteries and the superficial ones which lie just under the skin. The superficial veins in the leg may become varicosed when the valves are faulty and the veins enlarge. The superficial veins in the arm lie very near the surface at the bend of the elbow and are used for intravenous injections and transfusions.

The pulmonary circulation
There is another set of vessels which carries the blood to and from the lungs. This is called the pulmonary circulation and its function is to carry venous blood containing carbon dioxide to the lungs for excretion and to bring oxygenated blood back to the heart. It consists of the **pulmonary artery** which leaves the right ventricle, dividing to send a branch to each lung. Blood returns to the heart from the lungs by four **pulmonary veins**. These veins enter the left atrium.

The portal circulation
A fourth circulation carries blood, which is rich in food substances, to the liver from the stomach, spleen and bowel. This circulation consists of veins from these organs which join to form the large **portal vein** which enters the liver (Fig. 6.7).

THE BLOOD
The blood is a red sticky fluid and salty to taste. It is also warm and you should have learned these facts from your own observations. You will, however, see blood being withdrawn from the veins of patients, mixed with a chemical and allowed to stand in a narrow tube until a sediment appears at the bottom. This sediment consists of the solid particles of the blood. These normally float in the fluid which you can see above the sediment. The solid part of the blood consists of the blood cells and the fluid part is the plasma.

The plasma
This fluid is mainly water but has certain important substances dissolved in it. Some of these substances come from the foods we eat such as:

Glucose—from **carbohydrate** foods such as potatoes, sugar and flour, and carried in the plasma to the muscles.

Amino-acids—from **protein** foods such as meat and fish, and carried to all the cells.

Fat—from butter and milk and carried to the fatty tissues.

Salts—such as sodium chloride (common salt), potassium and calcium (from milk). The calcium is carried in the plasma to the bones.

Vitamins—from all foods carried to all parts of the body.

Other dissolved substances are:

Waste products—*carbon dioxide* from the muscles and *urea* from the liver.

Albumin.—This makes the blood sticky.

Clotting substances.—These make the blood form a solid mass when it is shed and so prevents haemorrhage.

Antibodies—which protect against infection.

Hormones—the secretion of the glands.

The blood cells
There are three types of blood cells:
- Red blood corpuscles.
- White blood corpuscles.
- Platelets.

Red blood corpuscles
These are the most numerous; you will remember they come from the red bone marrow. They are made of a substance called **haemoglobin** which, when it gets to the lungs, snatches the oxygen out of the air. It then carries the oxygen round the body to all the cells, but of course a particularly large supply goes to the muscles. People who are anaemic usually have abnormal red corpuscles and quickly become tired because the muscles cannot get an adequate amount of oxygen. The commonest type of anaemia is one where the haemoglobin does not have enough iron and this can be remedied by taking an iron tonic. Haemoglobin is a compound of **iron** and a protein but these substances, got from food, cannot form mature red blood corpuscles unless vitamin B_{12} is present. If this is not present the patient suffers from a more serious type of anaemia called pernicious anaemia. These red blood cells are sometimes called **erythrocytes** and have a limited life only lasting about 4 months. They are destroyed in the spleen and are replaced by the bone marrow.

White blood corpuscles
These are the cells which fight infection. Some of them, **leucocytes**, like the mobile army going abroad to fight the enemy on their own ground, can squeeze through the walls of the capillaries into the tissues to fight the invading bacteria. The others, **lymphocytes**, like the Home Guard wait at home to destroy the bacteria which manage to evade the rest and enter the blood stream. In some diseases and as a side effect of some drugs, the number of white blood corpuscles falls dangerously low with resulting increased risk of infection.

Platelets
These are very small cells which help in the clotting of blood. Blood will clot when it has escaped from a ruptured blood vessel. This is nature's defence against haemorrhage. A network of fine fibres appears in the blood, the cells get caught up in this and the resulting clot forms a plug for the bleeding vessel. A clot will also form inside a vessel when the lining has been damaged by disease or injury. This is a thrombosis.

If you want to speed up the rate at which the clot is forming in a bleeding wound apply a warm, wet, rough dressing; never a smooth oily one.

Blood withdrawn for transfusion or for certain tests has sodium citrate added to it because this chemical prevents the formation of the clot.

Blood groups

When a patient is to have a blood transfusion you will notice that the doctor takes a specimen of the patient's blood. This is sent to the Haematology Department to be 'grouped and cross-matched'. Before blood is transfused it must be checked very carefully to ensure that the patient is given blood which is **compatible** with his own. He must be given blood of the correct group.

What do we mean by blood groups? In our red blood cells we have some protein materials called A and B. Some of us have only A, we are said to belong to **Group A**. Those with only protein B belong to **Group B**. If both proteins are present the individual belongs to **Group A,B**. and if none are present he belongs to **blood Group O**.

In our blood plasma we have other protein materials which will react in the presence of the corresponding protein in the blood cells. If a person belongs to Group A he will have an Anti-B protein in his plasma. If he is transfused with Group B blood the cells will clump together. This blood is said to be incompatible and the patient's temperature may rise, he may become jaundiced and the clumped cells may block the kidney tubules and cause renal failure.

In the blood of more than three-quarters of the world population there is a quite different substance called the **rhesus factor** so named because it was first demonstrated in the rhesus monkey. Those of us who have this factor are said to be rhesus positive (rh + ve). If it is absent we are rhesus negative.

If rhesus positive blood is given to a rhesus negative individual there may be no reaction with the first transfusion but subsequent transfusion will be dangerous. A pregnant woman who is rhesus negative may be carrying a foetus which is rhesus positive. A reaction may occur in the blood cells of the foetus causing destruction of red cells and jaundice. To prevent severe damage to the child it is sometimes necessary to change completely the child's blood before or after birth.

You can see now how important it is to check both the blood group and the rhesus factor before starting a new blood transfusion.

HOW THE CIRCULATION WORKS

Let us now consider the work of this system as a whole. The piping is all connected up to the pump and the whole apparatus is full of blood, almost six litres in all. The heart pumps once every four-fifths of a second; normally about seventy-five times per minute. This pumping action is brought about by the contraction of the myocardium which pushes the blood out into the arteries; this is followed immediately by a period of rest when the whole heart dilates. During this phase blood is sucked into the right and left atria from the inferior and superior venae cavae and the pulmonary veins (Fig. 4.4). It then passes through the tricuspid and mitral valves into the ventricles. The heart then contracts, the valves close and the blood is pushed out of the left ventricle into the aorta and from the right ventricle into the pulmonary artery.

The septum not only divides the heart into two sides but separates the oxygenated or arterial blood from the venous blood. The venous blood is bluish-red in colour and contains the waste product, carbon dioxide, which is being carried to the lungs for excretion. As the heart dilates this blood is sucked into the right atrium through the venae cavae. Then it is pushed by the contraction of the heart muscle up the pulmonary artery into the lungs where the carbon dioxide is exchanged for oxygen. This oxygenated blood now returns by the four pulmonary veins to the left atrium and from there into the left ventricle which pushes it into the aorta. This is the beginning of the general circulation.

We hear a lot about 'holes in hearts' and 'blue babies'. These conditions may be due to a defect in the septum which allows some of the venous blood to mix with the arterial blood instead of going to the lungs. This accounts for the blue appearance or cyanosis of these patients and for their breathlessness and inability to walk far or to run. Remember that muscles require oxygen. This should be supplied by the arterial blood, but if this blood has become mixed with venous blood (rather like the sewage getting into the water supply), it is no longer able to carry the correct amount of oxygen to the muscles.

HOW THE CELLS GET THEIR NOURISHMENT AND OXYGEN

All the body cells require food and oxygen. These substances are taken from the tissue fluid which surrounds them. **Tissue fluid** is part of the blood plasma which has been pushed through the fine porous walls of the capillaries. It then circulates through the tissues giving up its oxygen and food substances to the cells and collecting carbon dioxide and other waste products. The tissue fluid is then sucked back into the blood and lymphatic capillaries (Fig. 4.6). The substance responsible for the 'sucking' is the plasma protein (albumen). If there is a loss of this substance, or an abnormality of the lymphatic vessels, tissue fluid collects in the tissues and **oedema** results.

BLOOD PRESSURE

This is the force which pushes the fluid through the walls of the capillaries therefore it is important that it is maintained within normal limits. The greatest force is required to take the oxygen to the brain cells and they will be the first to suffer as a result of any fall in blood pressure. This fall will also slow the circulation of the blood through the skin and make the patient more prone to pressure sores. Any great rise in blood pressure may result in the capillaries in the brain rupturing and the resulting blood clot damaging the brain tissue—a cerebral haemorrhage.

There are several factors which, working together, maintain the blood pressure within the normal limits. These are:

The amount of blood circulating.
The force with which the heart pumps.
The elasticity of the large arteries.
The state of the small blood vessels in the skin.

Shock and haemorrhage will lessen the amount of blood circulating, therefore the blood pressure of a badly injured patient or one who has had a major operation will fall. Blood transfusions are given to combat this and these patients may have their blood pressure recorded every quarter of an hour. This is done by using an instrument called a **sphygmomanometer**.

A weak or feeble heart beat also lowers the blood pressure and the loss of elasticity of the arteries raises it. In old age the elastic wall of the large arteries may become hardened; this is arteriosclerosis and means that they no longer stretch when the heart pumps blood into them so that the pressure exerted on the walls is greater. This increase in blood pressure may not affect the large arteries but may actually rupture the capillaries. This is a common cause of cerebral haemorrhage or stroke.

Small arteries in the skin dilate and contract with heat and cold, working like a thermostat keeping the body temperature normal. As they dilate the pressure of the blood decreases, therefore heat lowers the blood pressure and cold raises it. When the blood pressure falls the brain cells become less active because they are not getting their usual supply of oxygen and glucose. This accounts for the fact that hot weather and hot baths make us feel sleepy. The night nurse should remember the beneficial effects of a hot bath for a sleepless patient. The badly shocked patient may feel cold but it is dangerous to overheat him as dilating the small arteries may make his skin feel warmer but will still further lower his blood pressure.

THE RETURN OF THE BLOOD TO THE HEART

The cells get their oxygen because of blood pressure but it is mainly suction which carries the carbon dioxide away. As the heart dilates between each contraction venous blood is sucked into the right atrium. This is assisted by the action of the diaphragm in breathing and by muscle action and gravity. Whenever possible patients should be encouraged to move around in bed if they cannot be got up. Deep breathing exercises are also helpful in speeding up circulation.

The action of muscles on the veins is most important in the lower limbs where the blood must travel against gravity. The veins of the legs contain valves which prevent the back-flow of blood. These valves are faulty in varicosed veins and the veins become swollen and engorged with blood. The limbs swell and are painful; the skin is incompletely nourished and ulcers may develop.

Standing for a long time without moving will slow down the venous return, lower the blood pressure, and may result in fainting.

A patient may not be able to move his limbs as a result of pain following an operation and of the application of a plaster of Paris cast. The affected limb must be elevated to encourage the blood to return by gravity. As soon as possible the patient is encouraged to move his fingers or toes and the resulting muscle action helps to return the blood to the heart. The improved circulation decreases the swelling and helps to relieve the pain.

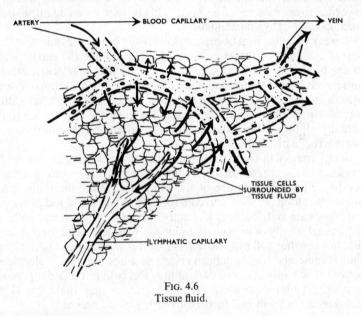

FIG. 4.6
Tissue fluid.

THE CARDIAC CYCLE

When you learn to measure blood pressure you will notice that two pressures are recorded, a high pressure and a low pressure.

The higher pressure is normally over 110 millimetres of mercury (110 mm/Hg), and is the pressure of the blood in the arteries when the heart is contracted. This is called **systole** and the pressure is **systolic blood pressure**. Systole lasts for 2/5 of a second. This contraction is followed by a period of rest when the heart is dilated and blood is being sucked back into the atria. This period also lasts 2/5 of a second and is called **diastole**. During diastole the heart is not pushing the blood out into the arteries therefore the pressure is less, usually around 80 mm/Hg. This is called **diastolic blood pressure**.

The **cardiac cycle** consists of systole plus diastole and takes place every 4/5 of a second in the normal heart.

THE PULSE

This is the beat of the heart as felt at an artery. The pulsation is caused by the stretching of the elastic walls of the arteries as the heart contracts and pushes the blood into them. By counting the pulsations we can record the rate at which the heart is beating but the nurse must never consider that this is enough because the feel of the artery under her fingers will tell her many things about the state of the circulation. The artery may feel soft or hard and wiry and the pulse weak or full and bounding. These tell her about the state of the artery wall, the strength of the heart and the amount of the blood circulating. A weak rapid pulse may indicate haemorrhage and is a dangerous sign. Irregularity in the rhythm of the pulse denotes a correspondingly irregular heart beat.

The Lymphatic System

The lymphatic system consists of an additional set of vessels through which some of the tissue fluid passes before reaching the large veins and entering the blood. These vessels consist of lymphatic **capillaries** and lymphatic **vessels** which lie in practically every tissue of the body. They join up with each other and eventually join two larger vessels or **ducts** which enter the large veins of the neck (Fig. 4.7).

The tissue fluid which finds its way into these vessels is called **lymph**. It is similar to the plasma of the blood but in cases of infection may contain bacteria. These bacteria, except in cases of severe infection, are prevented from entering the blood stream because the lymphatic vessels enter numerous small **nodes** containing white blood corpuscles. Infected lymph causes these nodes to swell and to produce more lymphocytes to fight the infection, thus the lymphatic system is one of nature's defences against disease.

Remember that the white blood corpuscles are of two types: the mobile army and the home guard. There are many occasions when the army of white blood corpuscles is defeated on enemy territory (infected tissue) and the invading organisms enter the lymphatic vessels. During the war we prepared for invasion by building small forts manned by the Home Guard at important road junctions. The lymphatic nodes are nature's forts, manned by the lymphocytes and situated at strategic points so that all the lymph must pass through them before entering the blood stream. Infection from a sore throat may result in a battle in the nodes in the neck, they become swollen and painful and can be palpated easily. These nodes are called the

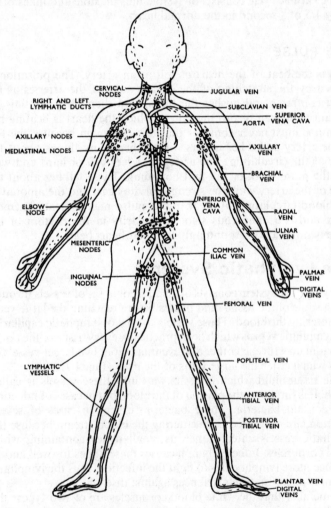

Fig. 4.7
The lymphatic nodes and the veins.

cervical lymph nodes sometimes referred to by the doctor as 'glands'. A septic finger may produce swelling of the **axillary** nodes in the axilla and a septic toe swelling of the **inguinal** nodes in the groin.

There are also lymphatic nodes in the pelvis, abdomen and thorax. The abdominal nodes are situated in the **mesentery** which is the covering of the bowel, and the thoracic nodes in the **mediastinum** which is the space between the lungs.

The lymph having travelled through these nodes now reaches the blood stream through two ducts which enter the subclavian veins in the neck. There is a short **right lymphatic duct** but the left duct is much longer as it drains the lymph from all the body below the diaphragm. It starts in the abdomen and passes up the thorax behind the heart. It is called the **thoracic duct**.

The tonsils and spleen

The tonsils and the spleen are made of similar tissue to the lymphatic nodes and therefore have a protective function. The tonsils guard the throat. The spleen, a purplish, half-moon-shaped organ, situated at the upper left hand corner of the abdomen becomes enlarged in acute infections and produces more white blood corpuscles.

Until about twenty-five years ago Man had to rely on Nature, and Medicine could do little to prevent the spread of infection in the body. Poultices and fomentations temporarily augmented the army of white blood corpuscles but nothing more. Now Man has devised a method of dive-bombing the invading organisms in the form of chemotherapy and death from blood poisoning following a septic finger is becoming a thing of the past.

Questions

(1) The organs of the circulatory system are the .... and the ....
(2) The heart lies in the .... cavity.
(3) The upper chambers of the heart are: (a) ventricles (b) septa (c) atria.
(4) The muscular wall of the heart is: (a) endocardium (b) pericardium (c) myocardium.
(5) The covering of the heart is the ....
(6) The valve in the left side of the heart is the .... valve.
(7) The word that means too much fluid is collecting in the tissues is ....
(8) The blood vessels are lined with: (a) epithelium (b) muscle (c) fibrous tissue.

(9) The blood supply of the heart is the .... circulation.
(10) The vessels returning blood to the heart from the general circulation are the .... and ....
(11) The blood leaves the left ventricle by the: (a) aorta (b) pulmonary artery (c) vena cava.
(12) The right atrium: (a) sends impure blood to the lungs (b) receives pure blood from the lungs (c) receives impure blood from the body (d) sends blood to the general circulation.
(13) The fluid part of the blood is called the ....
(14) The red blood corpuscles are made of ....
(15) The red blood corpuscles: (a) protects against infection (b) help in the clotting of blood (c) carry oxygen.
(16) There are two types of white blood corpuscles: (a) .... (b) ....
(17) Blood will not clot if it has had .... added to it.
(18) Tissues fluid passes through the walls of the: (a) veins (b) capillaries (c) arteries.
(19) The force of the blood against the artery walls is called: (a) pulse (b) blood pressure (c) circulation.
(20) The instrument used for taking blood pressure is a ....
(21) One of the following lowers the blood pressure. Which one? (a) a cold bath (b) shock (c) hardening of the arteries (d) excitement.
(22) One of the following raises the blood pressure. Which one? (a) a cold bath (b) haemorrhage (c) shock (d) a hot bath.
(23) One of the following helps the blood to return to the heart: (a) movement (b) valves (c) the force of the heart.
(24) In haemorrhage one would expect the pulse to be .... and ....
(25) The lymphatic system: (a) carries oxygen (b) helps in the clotting of blood (c) produces white blood corpuscles.
(26) Swollen axillary lymph nodes may be present in sepsis affecting ....
(27) The spleen lies in the: (a) right side of the abdomen (b) the pelvis (c) the left side of the abdomen (d) behind the heart.
(28) The function of the tonsils is to ....

5. The Respiratory System

We have seen that the cells of the various tissues require oxygen and that this is carried to them by the red blood corpuscles. We have also learned that after using up the oxygen the cells give off carbon dioxide. This interchange of gases is known as **respiration**; and in order that it should occur in the tissues, it must also occur in the lungs.

The **lungs** lie in the thorax and are connected to the air outside by a complex series of passages. These **air passages** allow the fresh air to enter the lungs where it comes in close contact with the blood; this is **inspiration**. The blood takes up some of the oxygen from the air and gives back carbon dioxide. As a result, the air breathed out contains more carbon dioxide and less oxygen. This air is called expired air and the act of breathing it out, **expiration**.

THE AIR PASSAGES

These have hard bony or cartilaginous walls, so that they can resist pressure from outside. If the walls were soft, like those of a rubber catheter, outside pressure could squeeze the walls together and so obstruct the passage of air.

These air passages are lined with delicate **ciliated** mucous membrane which warms and moistens the inspired air. This membrane produces a sticky mucus, to which particles of dust and soot adhere. Tiny hair-like processes called cilia projecting from the membranous lining keep the mucus moving along into the throat, where it is swallowed. If we catch a cold, this lining becomes inflamed—more mucus is produced—we become aware of it, and call it catarrh.

If you look at Figure 5.1 you will see that this system of air passages looks like an inverted tree. It is, however, a hollow tree with a hollow trunk, branches and twigs. The air enters through the **nose**, this leads into the **pharynx** or throat, then into the **larynx** or voice box, down into the **trachea,** the **bronchi,** the **bronchial tubes** and finally the very smallest passages, the **bronchioles.**

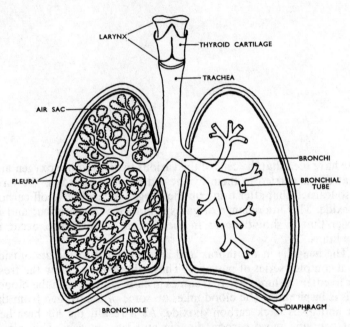

Fig. 5.1
Air passages showing bronchial tree.

What about the leaves of this bronchial tree? At the ends of the bronchioles there are small air-containing sacs. These resemble very tiny balloons—but not ordinary round balloons! These look more like raspberries from the outside. I am sure you would agree that much more rubber would be needed to make raspberry-shaped balloons than to make plain round ones, as their surface area is so much greater. The **air sacs** are made of fine porous epithelium and there are so many of them, and so much epithelium is used in their walls that if it were all spread out it would cover the floor of a very large room. This epithelium lies between the air in the sacs and the blood in the blood vessels and it is through this tissue that the oxygen and carbon dioxide are interchanged. Any thickening of this tissue will result in breathlessness.

The nose
This is a large cavity in the skull, divided into two by a septum. Its walls are bony except for the part which projects from the face; it is made of cartilage.

The Respiratory System

The air enters by the nostrils and is warmed, moistened and filtered by the mucous membrane. It then passes into the pharynx, except for a small amount which finds its way into the air sinuses and into the middle ear. The fact that these spaces in the skull communicate with the nose explains why infection can spread easily from the nose to the sinuses, causing sinusitis, and to the middle ear, causing otitis media.

Passing from the roof of the nose into the brain through tiny openings are the little hair-like endings of the **nerve of smell.** Smells are always in the form of gases. Some, usually the unpleasant ones, are so powerful that they have no difficulty in reaching the roof of the nose; others have to be sniffed up.

The pharynx

This is the throat. It is connected to the nose, the mouth, the larynx and the oesophagus rather like the junction of four busy streets. Fortunately there are 'traffic lights' and so when you breathe, you cannot swallow. If you try it, you will choke and cause a traffic jam! This is because, when food, mucus and saliva enter the throat and we swallow, a leaf-shaped piece of cartilage, the **epiglottis,** closes the larynx and directs the food into the oesophagus. But, if we take a breath when food is in the pharynx, the larynx will open and the food will be inhaled (Fig. 5.2).

We all know what happens when food goes down 'the wrong

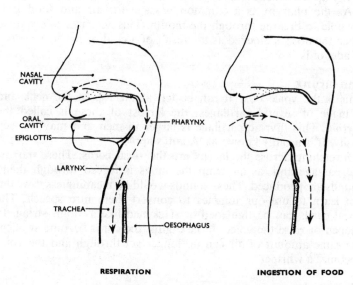

Fig. 5.2
The pharynx (diagrammatic section).

way' and we choke: coughing will dislodge the piece of food which has been inhaled into the air passage. But what happens when a patient is unconscious? This patient can neither swallow nor cough—the 'traffic lights' have stopped working! If food enters his pharynx, it may go in the wrong direction, blocking the air passages with possibly fatal results. Of course you would not be stupid enough to try to feed an unconscious patient by mouth. These patients are fed by passing a tube through the pharynx, into the oesophagus and down into the stomach. The mucus and saliva which will also block the airway must be sucked out by a sucker which acts in the same way as the little one used by the dentist, when he wants to dry up the mouth, before he does a filling.

Now think about the preparation of the patient before an anaesthetic is given. No food should be given during the preceding four hours because anaesthetics render the patient unconscious and temporarily paralyse his swallowing and cough reflexes. Many anaesthetic substances irritate the stomach and cause vomiting. If, therefore, there is any food in the stomach at the time, it will be vomited back into the pharynx and some of it may be inhaled into the lungs.

The conscious person is continually swallowing mucus and saliva. When a patient is anaesthetized these substances too, may trickle down into the lungs and block up the air spaces. To prevent this happening drugs such as atropine are given before operation to dry up the mucus and saliva.

As the pharynx is a common passage for air and food it is possible to breathe through the mouth. This acts as a safety valve when the nose is blocked as the result of a cold or by the presence of adenoids.

The larynx
This is the voice box. It can be felt at the top of the neck and is made of several cartilages; the largest of these is called the **thyroid**. This thyroid cartilage is larger in men and may be seen easily. It is usually known as 'Adam's apple'.

Stretching across the larynx are the **vocal cords**. These narrow the passage and, as air from the lungs is forced through them, sounds are produced. These sounds would be meaningless if we did not learn to use our tongues to convert them into speech. The vocal cords can be tightened or slackened like a violin string, to deepen or raise the voice. In laryngitis, the cords become swollen, the same amount of air can no longer get through and the voice becomes a whisper.

The trachea
This is the windpipe. It is made of rings of cartilage and can

be felt in front of the neck. It passes down into the thorax and divides to form the two **bronchi** which pass into the lungs.

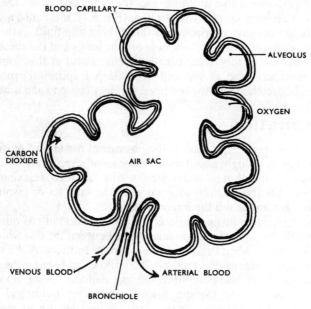

Fig. 5.3
An air sac.

The bronchial tubes

These are the small branches of the bronchi. They are made of rings of cartilage and muscle. These tubes get smaller and smaller until they become the muscular **bronchioles** which are the twigs of the bronchial tree. Each bronchiole ends in a group of **air sacs** (Fig. 5.3). Porous epithelium forms the walls or **alveoli** of the air sacs and these are surrounded by a network of blood capillaries. These capillaries are part of the pulmonary circulation (p. 59) and connect the pulmonary artery carrying blood rich in carbon dioxide to the pulmonary veins containing blood rich in oxygen.

THE LUNGS

These are cone-shaped, spongy organs situated in the thorax on either side of the heart. As the heart lies slightly over towards the left side the left lung is smaller than the right, having only two lobes whereas the right lung has three lobes. The base of each lung rests on the diaphragm and the apex extends up into the neck just behind the clavicle.

The lungs consist of the bronchial tubes, the bronchioles, the air sacs and the blood vessels, all held together by elastic tissue. On the outside they have a covering of serous membrane called the **pleura**. This membrane is like a double sac, the outer layer lines the chest wall and the inner layer covers the lung (Fig. 5.1). It is called a serous membrane because it produces a thin lubricating fluid, rather like serum, which prevents friction between the lungs and the chest wall. Inflammation of this tissue is called pleurisy and in this condition pain is experienced at the end of a deep inspiration when the pleura is stretched and the inflamed surfaces rub on each other.

Respiration

As we have seen respiration is the passage of air in and out of the lungs. This is brought about by the muscles of respiration increasing the size of the thorax when we breathe in, then relaxing, and returning the thorax to its original size. The muscles of respiration are the **diaphragm** and the **intercostals**.

The diaphragm forms the floor of the chest and flattens out when we breathe in. The intercostal muscles lie between the ribs which are raised upwards when these muscles contract. In this way the thorax is increased in size in all directions. This allows air to enter the lungs, and being elastic, they are stretched like balloons being blown up. As the ribs fall and the diaphragm returns to its normal dome shape, pressure is exerted on the elastic lungs; they recoil and air is expelled. There is, however, always some air left in the lungs, therefore they are never completely collapsed.

Other muscles play a part in deep or difficult breathing. These are the abdominal muscles and some of the muscles which move the shoulder girdle. Watch a patient who has had an abdominal operation, he may want to cough but will suppress it, because coughing is a deep expiration and means using his painful abdominal muscles. If he does not cough, mucus will collect in his bronchial tubes with serious results, so he must be encouraged to sit up and to support his abdominal wall with both hands while he coughs. You may have been taught to give your patient, who has difficulty in breathing, a bed or cardiac table. He can then lean forward, fix his shoulders and use the shoulder muscles to help the intercostals to raise the chest. This makes breathing a little easier.

It is a well-known fact that if breathing stops for more than 3 minutes permanent brain damage or even death will result. Try holding your breath and you will find that long before the 3 minutes are up you have involuntarily taken another breath. This is because there is a special centre in the medulla of the brain (p. 108) which is stimulated by the amount of carbon dioxide in the blood. When this reaches a certain level involuntary breathing occurs. Not only does the amount of carbon dioxide control respiration but it also controls the depth of respiration. The more exercise you do, the more carbon dioxide is produced in the muscles, the deeper you breathe.

RECORDING RESPIRATION

A normal adult breathes about sixteen to eighteen times per minute. This is counted by watching the number of times the chest rises or falls. Do not be content with recording the rate of respiration but notice the depth and where the movements are taking place. An unconscious patient's respirations may be so shallow that you cannot see much movement at all. You need not worry very much about this as long as the patient's colour is good, indicating that he is getting enough oxygen. If, however, his colour is poor, then even if he is taking deep breaths, there is some obstruction to the air passages. Either his tongue has fallen back into his pharynx or mucus is collecting there and blocking the entrance to the larynx. Immediate action must be taken to clear his airway or he may die.

Questions

(1) Respiration occurs in the lungs and in the ....
(2) The air passages are: (a)................................
   (b)........................ (c).....................
   (d)........................ (e).....................
   (f)........................ (g).....................
   (h)........................
(3) The air sacs are made of............................
(4) The nose is lined with..............................
(5) The air sinuses which communicate with the nose are in (a) the maxillary bone (b) the mandible (c) the ethmoid (d) the frontal bone. Which one is wrong? (Chap. 2.)
(6) The functions of the nose are (a) ....
   (b)................................
(7) The throat is: (a) the larynx (b) the pharynx (c) the phalanx.
(8) The pharynx is part of the respiratory and the ....
   .... systems.
(9) The large cartilage in the larynx is the ....
   cartilage.
(10) The larynx is the other name for the ....
(11) The bronchi are made of ....
(12) The tissue which holds the parts of the lungs together is:
   (a) cartilage (b) elastic (c) adipose (d) epithelium.
(13) The pleura is the .... of the lungs.
(14) On inspiration the diaphragm (a) falls (b) rises.

(15) The normal adult respiration rate is .... times per minute.
(16) The respiratory centre in the brain is stimulated by the amount of .... in the blood.
(17) Two things commonly block the airway of an unconscious patient. These are (a) .... (b) ....

6. The Digestive System

The Digestive System is a long tube which can be opened at either end. It extends from the mouth to the anus. It is made of muscle and resembles a 30-foot-long (10 metres) conveyor belt along which the food passes. At various points along this belt the food is sprayed and mixed with digestive juices. These juices contain chemical substances called **enzymes** which break down complex protein, carbohydrate and fatty foods into simple soluble substances. As a result, before it is halfway through, your last meal is completely unrecognizable: those potatoes, that rice pudding, the sugar in the coffee all look the same now. They have been pounded down and mixed up and also broken down chemically by the enzymes into a soluble sugar called **glucose**. The lamb chop and the green peas have also been changed into soluble forms; these are called **amino acids**. The fat of the chop, the butter on the bread and the cream in the coffee are changed into **glycerin** and **fatty acids**. These soluble substances can be absorbed; this means that they can pass from the digestive canal into the blood stream. Do you remember that plasma contained glucose and amino acids but no glycerin or fatty acids? This is because the glycerin and fatty acids are recombined into human fat before they reach the blood stream.

There are other substances in our food such as water, mineral salts and vitamins. These are already soluble and do not require to be digested. The hard part of fruit and vegetables is called cellulose; this cannot be broken down and used by the body; in other words it is indigestible and passes unchanged right through to the last part of the conveyor belt to form faeces. In our diet this cellulose is called roughage.

The digestive system may be described as consisting of two parts:
    The food or alimentary canal
    The digestive glands

The *Alimentary Canal* (Fig. 6.2) consists of the following parts:
    The **mouth**.

The **pharynx**.
The **oesophagus**.
The **stomach**.
The **small intestine**.
The **large intestine**.
The *Digestive Glands* are:
The **salivary glands**.
The **gastric glands**.
The **pancreas**.
The **intestinal glands**.
The **liver**.

The Alimentary Canal

The mouth

The *Mouth* is the beginning of the canal. It is a bony cavity with a muscular floor. It contains the teeth and the tongue and its function is mastication.

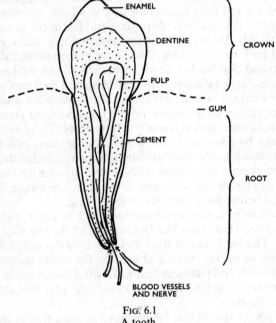

Fig. 6.1
A tooth.

The **teeth** perform the important mechanical breakdown of the food. The **incisors** and **canines** are the sharp teeth in front and they bite and cut. At the back are the larger flat teeth, the **molars** and

pre-molars, which grind and pulp the food. Every individual has two sets of teeth, the temporary set which are cut during the first three years of life and the permanent teeth which replace them from about the age of 6.

There are 32 permanent teeth each one of which has a crown, the part above the gum, and a root below the gum. They are made of a hard substance called dentine which is covered with enamel above the gum and cement below. Each tooth is hollow; this hollow contains a pulpy substance into which enters the blood vessel and nerves through a hole in the root (Fig. 6.1).

Entering the mouth are the ducts of the **salivary glands** which are situated around the mouth (Fig. 6.2). The **parotid glands** lie in front of the ears, the **sub-mandibular** and **sub-lingual** in the floor of the mouth. These glands secrete **saliva,** a watery fluid which helps to keep the mouth moist, softens the food and starts off digestion. These glands increase their secretion at the thought, the sight or the

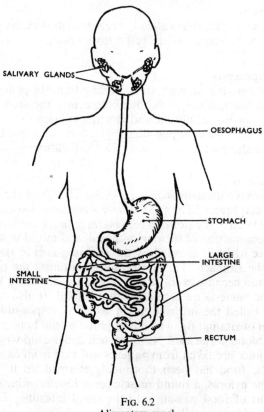

Fig. 6.2
Alimentary canal.

smell of food. This is a very good reason why we should have our meals about the same time each day and why we should serve the patient's meals punctually.

Infection from a dirty mouth may pass along the ducts into the glands and therefore an important part of a nurse's duty is to attend to the patient's oral hygiene.

The **tongue** lies in the floor of the mouth to which it is attached. It is a muscular organ covered with mucous membrane. This mucous membrane is rough in appearance because of the little projections called **papillae** which contain the taste buds. These buds are the endings of the **nerve of taste** and are situated all over the tongue; those at the tip and sides are stimulated by sweet, salt and sour and those at the back by bitter substances. The tongue helps to form the food into a soft ball called a bolus. This is the act of chewing or mastication. The tongue is also necessary for swallowing and speech.

The pharynx
The pharynx or throat has already been described in the preceding chapter as it is also part of the respiratory system.

The oesophagus
The oesophagus is a muscular tube leading from the pharynx down behind the heart, through the diaphragm into the stomach. It is lined with mucous membrane and this mucus helps the food to slip down. The food is also helped along by a contraction and relaxation of the muscular wall, a movement called **peristalsis**.

The stomach
The stomach is a muscular bag lying in the left side of the abdomen under the diaphragm. The food simply slips down the oesophagus but it is retained in the stomach for several hours. Here it is churned up by the contraction of the muscular wall and mixed with the acid **gastric juice** which is secreted by the **gastric glands** in the mucous lining of the stomach. This gastric juice continues the process of digestion and because of its acidity acts as a disinfectant killing off most of the bacteria we swallow with our food. It also contains a substance called the intrinsic factor which is responsible for the absorption of vitamin B_{12} which is required by the bone marrow to make red blood corpuscles. Now you will understand why samples of gastric juice are taken from patients suffering from anaemia.

After the food has been thoroughly churned up it is pushed towards the **pylorus**, a round muscle (or sphincter) which controls the amount of food passing into the small intestine. The partly digested food is now called chyme.

The small intestine

The small intestine or bowel is about 22 feet (7 metres) in length. It lies in the abdominal cavity like a great coiled tube held in place by a fold of the peritoneum, the serous lining of the abdomen. This fold, called the **mesentery,** carries the blood vessels and nerves out from the back of the abdominal wall to the bowel (Fig. 6.3).

The small intestine is muscular and lined with mucous membrane. This mucous membrane is thrown into folds to increase its surface area because it contains very important glands, the **intestinal glands.** These glands secrete the alkaline **intestinal juice** which

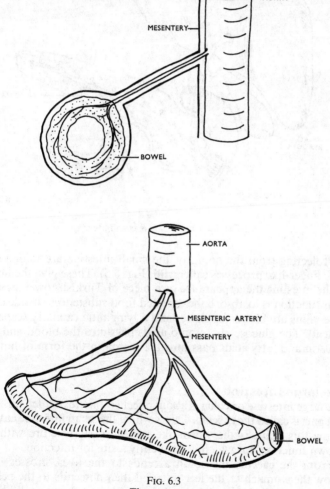

Fig. 6.3
The mesentery.

completes the digestion of the food substances into glucose, amino acids, glycerin and fatty acids.

The first part of the small intestine is called the **duodenum** (Fig. 6.4). It is a 'C' shaped loop which surrounds the head of the pancreas. Into the duodenum flow two more alkaline juices, **pancreatic juice** from the **pancreas** and **bile** from the **liver**. These mix with the food after it leaves the stomach and continue its digestion which will be completed by the intestinal juices.

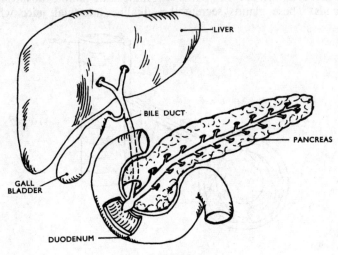

Fig. 6.4
The duodenum, pancreas and liver.

Projecting from the lining of the small intestine are millions of little finger-like processes called **villi** (Fig. 6.5). These give the lining of the intestine the appearance of a piece of Turkish towelling and their function is to absorb the digested food substances. Inside each little villus are blood capillaries and a lymphatic capillary (called a **lacteal**). The glucose and amino acids pass into the blood and the glycerin and fatty acids pass into the lymph in the form of human fat.

The large intestine

The large intestine or **colon** is about 5 feet ($1\frac{1}{2}$ metres) in length. The first part is called the **caecum**. This is where the small intestine enters. Below this point is the worm-like appendix, a structure with no known function but which is frequently a site for infection.

From the **caecum** the colon ascends to the liver, crosses over below the stomach to the left side and then descends to the pelvis. This part is called the **pelvic** or **sigmoid colon**; it forms an 'S' bend

The Digestive System

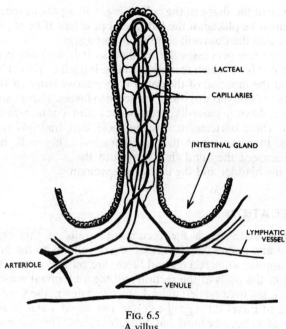

Fig. 6.5
A villus.

before it straightens out to form the **rectum**. The rectum passes straight down through the pelvis to the outside at the round muscle called the anal sphincter or **anus**.

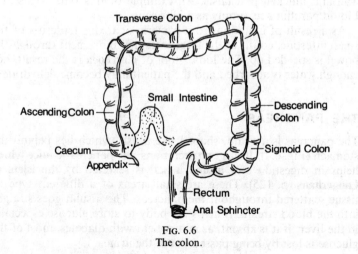

Fig. 6.6
The colon.

Because of the shape of the colon (Fig. 6.6) a patient receiving an enema must be placed in the left lateral position. If he is placed on the right side the fluid will simply run out again.

The digestive processes are finished so it is only the residue or waste of our food which reaches the large intestine. This is in a fluid form and the function of the colon is to remove some of the water and salts and convert this waste matter into **faeces**. Faeces are brown in colour, have a paste-like consistency and contain millions of bacteria. These bacteria, mainly the bacilli coli, multiply rapidly in the large intestine where they are harmless. They will, however, cause disease if they find their way into the other pelvic organs, namely the bladder and the uterus in a woman.

DEFAECATION

When the faeces reach the rectum the walls of this organ are stretched and nerves convey the feeling of fullness to the brain. On defaecation the anus relaxes and faeces are pushed out by the contraction of the walls of the rectum and the abdominal muscles.

Sometimes, because peristalsis has slowed down, the residue takes too long to travel through the colon, too much water is absorbed and the faeces become hard and difficult to pass. This is **constipation**. Constipation is best treated by taking a diet rich in fruit and vegetables, plenty of water to drink and when possible exercise. Sometimes, however, we may have to resort to the use of drugs. These drugs are called **aperients** and there are three different types. The saline aperients (Epsom salts) draw water back out of the walls of the colon and make the faeces fluid again. Others act on the muscular wall and quicken peristalsis. An example of this type is cascara. Liquid paraffin acts simply as a lubricant.

As a result of irritation from bacteria or toxins (enteritis of the small intestine, colitis of the large intestine) movement through the bowel is speeded up. The loose stool of **diarrhoea** is the result, not enough water is absorbed and the patient may become dehydrated.

THE PANCREAS

The pancreas is a gland, shaped like a fish, which lies behind the stomach (Fig. 6.4). It has two secretions, the pancreatic juice which helps in digestion and **insulin** which is secreted by the **islets of Langerhans** (p. 125). These are small areas of a different type of tissue scattered throughout the pancreas. The insulin goes straight into the blood stream. It helps the body to store glucose especially in the liver. If it is absent, as in a patient with diabetes, most of the glucose is lost by being passed out in the urine.

THE LIVER

This is the largest gland in the body (Fig. 6.4). It lies in the upper right hand side of the abdomen under the diaphragm. It is a wedge-shaped organ, soft, reddish-brown in colour and has many important functions.

The liver manufactures **bile**. This greenish-yellow fluid is stored in the **gall bladder** which lies on the under side of the liver (Fig. 6.4).

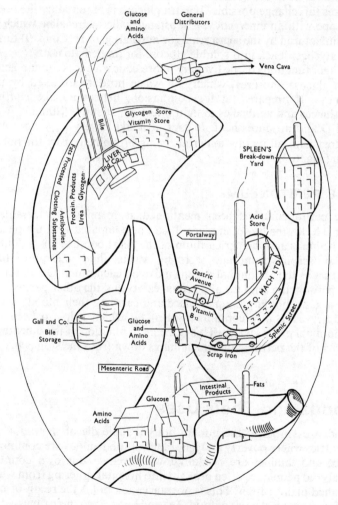

FIG. 6.7
'The Portal Industrial Estate.'
Diagram to illustrate the portal circulation and the functions of the liver.

The bile passes down the bile duct into the duodenum. Here it helps in the digestion of fat and it is the bile in the faeces which colours and deodorizes them. A patient becomes jaundiced or yellow when, for some reason or other, the bile duct becomes blocked. The bile gets into the blood, colours the skin, and the faeces having no bile are pale, fatty and foul smelling. The bile is excreted by the kidneys and the urine is a dark orange colour.

The liver also stores **glucose** in the form of **glycogen** which is insoluble. It is the action of the **insulin** from the pancreas which makes this change possible. If extra glucose is required by the body to cope with an emergency, a substance called **adrenaline** which is manufactured by the **supra renal** (or **adrenal)** glands (Chap. 9) turns the glycogen back into soluble glucose for immediate use.

Other functions of the liver are to break down excess amino acids into glucose and **urea,** which is a waste produce excreted by the kidneys. It prepares fat for combusion, helps to make **clotting substances** and **antibodies** to fight infection. It stores vitamins A, B and D and produces heat. In fact, it is like a very large chemical factory with all its raw materials being brought to it by the portal vein (Fig. 6.7).

THE PERITONEUM

The peritoneum has been mentioned as forming the mesentery which holds the bowel in place. It can be compared with the pleura of the lungs and the pericardium of the heart because it is a double sac of serous membrane secreting serous fluid and preventing friction as the abdominal organs move on each other. One layer of the peritoneum lines the abdominal cavity and the other part covers the organs, holding them in place and carrying their blood vessels and nerves. It also contains a lot of fat which helps to keep the abdominal organs warm. This is the main function of the omentum, a fold of the peritoneum which hangs down from the stomach.

Foods and Their Uses

Food is essential to the human body. People die of starvation in countries where poverty and ignorance of food values are commonplace and famines are not unknown. Even in our own country lonely old people are often brought into hospital suffering from what is called malnutrition. This is not starvation but is the result of not taking the correct types of food. Tea and buns taken three times a day is rather like filling the petrol tank of a car with water and expecting it to go!

We must have carbohydrates and fats to provide the fuel to keep us going; proteins, mineral salts and vitamins to replace and repair worn out parts; water to replace the amount which is being lost continually and roughage to keep the conveyor belt in action.

CARBOHYDRATES

The foods we require for **energy** and **heat** are called carbohydrates because they contain carbon, hydrogen and oxygen. Some of these carbohydrates are very insoluble and require a lot of digesting before they can be absorbed into the blood stream and used as a fuel. These are the starchy foods, potatoes and all foods made from flour. Sugars such as cane and beet sugar and the sugar of milk and malt are less insoluble and therefore more easily digested. All the carbohydrate foods, **starches** and **sugars,** must be converted into **glucose** by the digestive juices before they can be of any use to the body. Glucose can be taken in food, especially fruit drinks. It requires no digestion and can be instantly absorbed by the stomach. Glucose can also be given by intravenous infusion.

FATS

Fatty foods are rather like the carbohydrates both in their composition and their action. They provide **energy** and a great amount of **heat.** They also form food stores, the adipose tissue of the body, and protective coverings for some organs. Butter, cream, egg yolk, the fat of meat and fish, nuts and oils are the source of fat in our diet.

PROTEINS

Proteins, in addition to carbon, hydrogen and oxygen, contain nitrogen. They are called, therefore, the nitrogenous foods and are required to **build** and **replace** the protoplasm of our body cells. The best type of protein is obtained from animal flesh, meat and fish. Milk, egg white and cheese are also good sources and some vegetables such as peas and beans provide a less useful type of protein. All are broken down by the digestive juices into **amino acids** before they can be absorbed and utilized.

WATER

Water is absolutely essential for life as between 60 to 70 per cent. of our body weight is water. It forms all the body fluids such as blood, tissue fluid and lymph and is necessary for the formation of the secretions of all our glands. It is obtained from all our foods and is made in the body when carbohydrates and fats are burned.

MINERAL SALTS

Different tissues require different minerals. We have already learned that red blood corpuscles require iron, and bone and teeth calcium. All the body fluids require sodium chloride and potassium. The thyroid gland must have a supply of iodine. These mineral salts are obtained in our food from the following sources.

Sodium chloride is common salt and is present in most foods.
Potassium is present in milk and eggs.
Iodine is present in sea foods and in table salt.
Iron is obtained from vegetables, especially spinach, egg yolk and meat.
Calcium is found in milk, cheese, green vegetables and egg yolk.

VITAMINS

These are chemical substances found in certain foods and are necessary for perfect health. Their presence was only discovered in 1912 and many vitamin deficiency diseases, such as rickets, are now much less common. As each vitamin is discovered it is given a letter of the alphabet, thus vitamins A, B, C, D, E, and K.

Vitamin A

This vitamin is found in all fatty foods, milk, cheese, butter, liver and fish liver oils. It can be made in the body from a substance called carotene which is present in brightly coloured vegetables such as carrots, tomatoes and the outer leaves of cabbage and lettuce. This vitamin keeps our mucous membranes and our eyes in a healthy condition by helping to prevent infection. It also prevents **night blindness** and if your diet contains an adequate supply you will find it fairly easy to see where you are when you get off the bus in the dark. Cod liver oil is a valuable source of vitamins A and D.

Vitamin B

This is a very complex vitamin consisting of several different chemical substances. Vitamins B_1 and B_2 are present in the husk of wheat, oatmeal, whole rice, fruit, nuts and yeast. Deficiency affects the nerves and in countries where polished rice is the staple diet a disease called **beri-beri** is common and the patients suffer from painful neuritis.

Vitamin B_{12} is found in animal foods, especially liver. It is absorbed in the stomach, carried to the liver by the portal vein for storage. This vitamin is required for the normal development of the red blood corpuscles in the bone marrow, its absence is the cause of pernicious anaemia.

Most of this group of vitamins are built up in our intestines by the micro-organisms which normally live there. If disease-producing germs invade the body we try to kill them with substances called antibiotics. Some of these antibiotics will also kill the friendly germs so we must give our patients vitamin B tablets to replace the vitamins no longer being manufactured.

Vitamin C
This vitamin is very important and is given to young babies in orange juice. It keeps the teeth, gums and blood vessels healthy and helps in the **healing of wounds**. In addition to oranges it is present in lemons, black currants, rose hips, tomatoes and green vegetables. As this vitamin is easily destroyed by cooking these foods should be eaten raw whenever possible. Absence of vitamin C causes **scurvy**, a condition which used to be common in sailors on long sea voyages because of the lack of fresh fruit and vegetables.

Vitamin D
This vitamin is present in fatty foods, particularly fatty fish. It is found along with vitamin A in salmon, cod, herring and sardines and also in milk, cheese and butter. It can also be built up in our bodies. The ultra-violet rays from the sun act on a fatty substance in the skin and produce vitamin D. For this reason exposure of the skin to the sun's rays is beneficial, especially during the growing period, because calcium cannot be absorbed unless this vitamin is present.

If vitamin D is absent in a child's food he will suffer from **rickets**. His bones will become soft and pliable. Most mothers give their growing children extra vitamin D in the winter time in the form of cod liver oil and rickets is now practically unknown in this country.

Vitamin E
This vitamin is found in egg yolk, milk and green vegetables. It is thought to be necessary for reproduction.

Vitamin K
Vitamin K is found in green vegetables and is necessary for the clotting of blood.

Metabolism
Carbohydrates, proteins and fats can all be burned up for energy. This is metabolism, and metabolic rate is the rate at which we use them. In some conditions, such as goitre, the metabolic rate is greatly increased and these patients are so active that, no matter how much they eat, they remain thin.

Carbohydrates are burned up with oxygen in the muscles to release energy, but first of all they must be digested into soluble **glucose**. This glucose is absorbed through the walls of the villi in the small intestine into the blood stream. It is carried to the liver by a large vein called the portal vein. The liver stores some and sends the rest out into the general circulation and hence to the muscles. In the muscles the glucose and fat are burned up with oxygen, energy is released and **carbon dioxide** and **water** are given off. This cannot happen, however, unless there is **insulin** present in the blood stream.

Fats are burned up in the muscles with glucose and are also used by the body for protective purposes and as a food store. They are digested, absorbed into the lymphatic vessels, carried up the lymphatic duct into the blood stream and then circulated round the body to all the **adipose tissue**. The fat required by the muscles is carried to the liver first and made into a fit state for combustion. A patient with severe **diabetes** loses a lot of weight. He is losing glucose in his urine and because of this, large quantities of fat are broken down into glucose. This provides him with a source of energy but at the same time **ketone** bodies are accumulating causing coma and death if untreated. Acetone is a ketone body and this is why you must test a diabetic patient's urine for acetone as well as sugar.

Protein foods are used by the body for building new cells and repairing old or broken down ones. This is easily understood when we remember that the protoplasm of the cells is a form of protein and that most of our protein foods consist of the flesh of other animals. Protein is digested into **amino acids** and these are absorbed into the blood stream, then carried by the portal vein to the liver. The liver sends out what is required by the body into the blood. Any excess is broken up and converted into glucose and **urea**. The glucose is used as a fuel and the urea sent out to the kidneys where it is excreted in the urine. In destructive diseases, such as tuberculosis, a high protein diet is given in an attempt to balance the rate of repair with the rate of destruction.

Notice that carbohydrate, fat and protein can all be used as fuel releasing energy and heat. The amount of heat produced by each is measured in **calories**. A calorie is the amount of heat required to raise one kilogramme of water one degree Centigrade, for example, from 15° to 16°C.

One gramme of carbohydrate will produce 4 calories, 1 gramme of protein will produce 4 calories but 1 gramme of fat will produce 9 calories, so we get more heat from fatty foods.

The amount of food we require in 24 hours is calculated in calories and varies with age, sex, height and occupation. An average woman doing light work will require to eat enough food to produce 2,500 C. in 24 hours. An average man doing heavy work will require 3,500 C. Any food eaten in excess of this requirement will be laid down as fat.

Table 6.1
What Happens to Your Food
Ingestion, Digestion and Absorption

Organ	Movement	Juice—Glands	Action	Result
Mouth	Mastication (Chewing)	Saliva—Salivary Parotid Submandibular Sublingual	Starts digestion	Bolus
Pharynx	Swallowing	—	—	—
Oesophagus	Peristalsis	—	—	—
Stomach	Churning	Gastric—Gastric	Continues digestion Acidifies food and vitamin B_{12} is absorbed	Chyme (Semi-solid)
Duodenum ↓ Small Intestine	Peristalsis Peristalsis	Pancreatic—Pancreas Bile—Liver Intestinal—Intestinal	Continues digestion and emulsifies fat Completes digestion— | End products absorbed	Chyme End products Glucose Amino-acids Fatty acids and Glycerin and Waste (Fluid)
Large Intestine	Peristalsis	— —	Water absorbed Bacillus coli mixed with residue of food	Faeces (Paste)

Table 6.2
What Your Body Does with Your Food—Metabolism

Food	End product	Absorbed	Route	Liver	Action	Waste	Excreted by
Carbohydrate	Glucose	Through walls of villi in small intestine into blood capillaries	Portal vein to liver	Stored as Glycogen (*Insulin* required)	Carried to muscles—combustion with O_2 releases *heat and energy*	Carbon dioxide and water	Lungs Skin Kidney Bowel
Protein	Amino acids	Through walls of villi in small intestine into blood capillaries	Portal vein to liver	Excess broken down into—*Urea* and *Glycogen*	Carried to all tissues for *growth and repair*. Glycogen used for *heat and energy*	Urea Carbon dioxide and water	Kidneys Lungs Skin Kidney Bowel
Fat	Fatty acids and glycerin	Through walls of villi in small intestine into lymphatic capillaries (Lacteals)	Lymphatic system | Blood stream | Adipose tissue	Prepared for combustion	Combustion in muscles with glucose | *heat and energy* If no glucose (Diabetic)	Carbon dioxide and water Acetone	Lungs Skin Kidney Bowel Kidney

Questions

(1) The digestive system consists of two parts, the .... canal and the ....
(2) Which of the following is not a digestive gland? (a) sebaceous (b) gastric (c) salivary (d) liver.
(3) Which of the following is not a part of the digestive system: (a) the small intestine (b) the stomach (c) the larynx (d) the oesophagus?
(4) The .... and .... are the sharp teeth in the front of the mouth.
(5) The function of the molar teeth is to .... the food.
(6) The salivary glands are the: (a) ....
(b) .... (c) ....
(7) The tongue is made of ....
(8) The movement of the oesophagus is ....
(9) The pylorus lies between the .... and the ....
(10) The small intestine is about: (a) 6 feet (b) 22 feet (c) 32 feet (d) 7 metres (e) 2 metres—in length.
(11) The lining of the abdomen is called the ....
(12) The mesentery is part of the ....
(13) The mesentery supplies the bowel with .... and ....
(14) The .... is the first part of the small intestine.
(15) The pancreatic duct and the bile duct enter the ....
(16) Food is broken down by digestive juices from digestive glands. These are: (a) ....
(b) .... (c) ....
(d) ....
(17) The small intestine contains .... through which the food substances are absorbed.
(18) The anus is the: (a) posterior (b) anterior (c) middle— opening through the pelvic floor.
(19) The anus is a round .... called a ....
(20) The large intestine is called the ....
(21) The appendix is attached to the: (a) duodenum (b) large intestine (c) small intestine.
(22) The function of the large intestine is to convert the waste matter into ....
(23) This waste is excreted in the act of ....
(24) Constipation is caused by: (a) irritation of the bowel (b) lack

of roughage (c) slow peristalsis (d) lack of fluid—Which one is wrong?
(25) The pancreas has two secretions. These are .... and ....
(26) The liver stores: (a) bile (b) urea (c) fat (d) glucose.
(27) The livermanufactures: (a) bile (b) urea (c) fat (d) glucose.
(28) The bile helps to digest .... It is stored in the ....
(29) Carbohydrate foods are digested into: (a) amino acids (b) glycerin (c) glucose.
(30) There are two types of carbohydrate food (a) .... and (b) ....
(31) One of the following is a carbohydrate food. Which one? (a) fish (b) bacon (c) cereal (d) liver.
(32) Potatoes are .... foods.
(33) The vegetable source of fat in the diet is: (a) .... and (b) ....
(34) Carbohydrates and fats are similar in their composition but proteins contain ....
(35) Meat and fish are .... foods.
(36) Proteins are broken down into .... by the digestive juices.
(37) The mineral salt necessary for the formation of red blood cells is ....
(38) The thyroid gland requires ....
(39) The vitamins found in fatty foods are: (a) A (b) B (c) C (d) D.
(40) The vitamin most easily destroyed by cooking is vitamin: (a) A (b) B (c) C (d) D.
(41) Cod liver oil contains vitamins .... and ....
(42) Vitamin .... is necessary for the formation of haemoglobin.
(43) Lack of vitamin D in the diet can cause: (a) beri-beri (b) rickets (c) blindness.
(44) Carbohydrates and fatty foods are used for .... and ....
(45) The waste product of the metabolism of protein is: (a) water (b) urea (c) carbon dioxide.
(46) Insulin is necessary for the metabolism of ....

7. The Excretory System

Excretion means the passing out from the body of waste products produced as the result of body activity. These substances, if allowed to accumulate, would produce ill health and eventually death. Compare this with your knowledge of communal health. We could not live surrounded by our own excretions nor can the cells of our bodies. The wandering races, who live in tents, do not worry about what happens to their refuse; once its accumulation becomes unpleasant and unhealthy they just move on and pitch their tents elsewhere. We more civilized people, who live in permanent houses, have had to discover some means of disposing of our excreta. You will learn all about this sewage in your health lectures. This chapter will deal with the organs responsible for the filtering and disposal of the waste which must not be allowed to collect in our bodies.

In Chapters 3 and 5 we dealt with the excretion of carbon dioxide by the lungs. In Chapter 6 we discovered that the waste from indigestible parts of food is formed into faeces and excreted by the bowel. Other waste substances, such as excess water, urea and salts, are excreted by the skin and the urinary system.

The Skin

This is not only a covering for the body but a very important organ with many functions. In addition to excreting waste products in the form of sweat, it has a protective function; acts as a thermostat controlling the body temperature, produces hair and nails and is an organ of sensation. It deserves to be well cared for and can only carry out all these functions if it is kept clean and covered with the correct type of clothing so that air can circulate round it.

The **protective function** of the skin is an interesting one. If we go far enough back in history we find that our ancestors had hard scaly skin like coats of armour. Look at the skin of fishes and reptiles. It is obviously protective in function. Your skin may appear soft and smooth but under the microscope it is seen to consist of layer upon layer of cells (Fig. 7.1). This is called the **epidermis** and is thickest on the soles of the feet where most pressure is taken. The outer layer of the epidermis consists of hard scaly cells forming a barrier to bacteria: man's deadliest enemy. Once the barrier is broken by injury or burns bacteria can enter and multiply. A good healthy skin is a tremendous help in

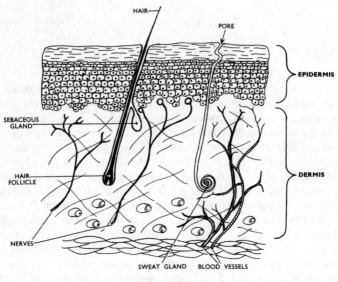

Fig. 7.1
The skin.

preventing infection. The top layer of cells of the epidermis is continually being shed. These discarded cells are washed off or removed with our clothes each night. Like all other dead matter they decompose and this is one of the reasons for the unpleasant smell of an unwashed person.

Because the skin contains special **nerve endings** which convey sensations of **touch, pain** and **temperature** to the brain, it helps to protect us against dangerous things in our environment. This is best explained by thinking of the difficulties met with in nursing a paraplegic patient. The messages from this patient's skin are blocked and never reach the brain, therefore it would be very easy to burn him with a hot water bottle as he does not feel the heat. He neither feels pressure nor pain and unless turned would lie quite comfortably in one position until a pressure sore has developed. On the other hand, this protective function is greatly increased when one of the other senses is missing. Watch a blind person 'seeing' with his hands and you will realize how important his skin is to him, especially the tips of his fingers where there are numerous nerve endings.

These nerve endings do not lie in the epidermis but deeper down in the **dermis** or true skin. This is elastic tissue allowing the skin to stretch with the movements of the body. Like the blood vessels, it becomes less elastic in old age and gets wrinkled.

Also in the dermis lie the **sweat glands**. These open out through the surface of the epidermis at a pore and produce **sweat**. This sweat consists of water and a little salt and is continually being excreted and evaporated into the surrounding air. As a result of exercise, excitement or too much heat these glands become more active and we become aware of this additional sweat coming to the surface and we say that we are 'perspiring'.

As well as getting rid of waste water and salts this evaporation of sweat helps to **regulate the body temperature**. This is a very important function of the skin. As you know, the temperature of the body is normally between 36° and 37°C. Various factors raise the temperature, exercise and additional heat from outside sources, such as the sun, fires and hot water bottles. In health this heat is balanced by an equal loss of heat. We lose heat by various routes. All excretions, faeces, urine and expired air are warm, but most heat is lost through the skin.

When you get hot you sweat and your skin gets red. This is because the dermis contains little muscular arteries which automatically dilate when you are warm, increasing the amount of blood circulating through the skin. This increases the loss of heat to the cooler air by radiation. At the same time the sweat glands produce more sweat to be evaporated. Evaporation is the turning of a liquid into a gas; this change requires heat and the heat is taken from the skin. There must also be air circulating before evaporation can take place and this is the reason why good ventilation and proper clothing are important. The more rapid the current of air the quicker the evaporation and the greater the heat lost, as you will soon discover if you sit in a draught.

In conditions where the body temperature rises above 39·4°C. some attempt must be made to lower it. This is done by tepid sponging. Tepid water is sponged on to the hot skin and allowed to evaporate. This uses up the extra heat and lowers the temperature. In more severe cases the patient may be covered by a wet sheet and an electric fan used to speed up the rate of evaporation.

Hair and **nails** grow out from the skin and, in addition to their 'cosmetic' function, the nails protect the tips of the fingers and toes and the hair helps to keep us warm by trapping air which is a bad conductor of heat. Hairs have their roots in the dermis and are present all over the body except on the soles of the feet and the palms of the hands. Near the root of each hair there is a small gland, the **sebaceous gland,** which produces an oily substance called **sebum.** This sebum keeps the skin **waterproof** and in good condition; it also lubricates the hair.

Sebum is a mild antiseptic but it is also greasy and there are millions of bacteria sticking to it. Because of this, some form of skin preparation must be performed before any surgical operation, otherwise the scalpel will carry the bacteria down into the deeper tissues and cause infection. This preparation consists of shaving the hair from the part and washing off the sebum with a detergent.

One other function of the skin has been mentioned in Chapter 2. It contains a substance **ergosterol** which can be acted upon by the ultra-violet rays of the sun to form **vitamin D.**

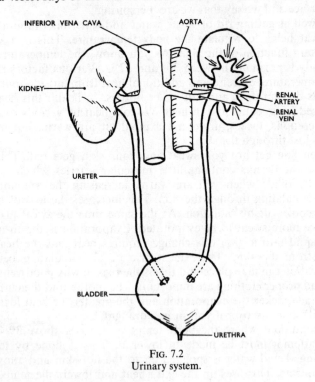

FIG. 7.2
Urinary system.

The Urinary System

This system consists of the following organs:
 The *Kidneys.*
 The *Ureters.*
 The *Bladder.*
 The *Urethra.*

THE KIDNEYS

These two bean-shaped organs lie on the posterior abdominal wall embedded in a protective pad of fat. Attached to each of them is a fine muscular tube, the ureter, which passes down into the pelvis to enter the bladder (Fig. 7.2).

Each kidney has a hollow central part. This is the upper expanded end of the ureter and is called the pelvis of the ureter. It resembles a funnel attached to tubing. The **solid portion** where the

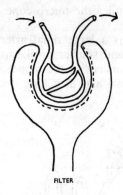

FILTER

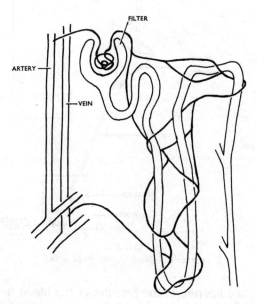

FIG. 7.3
A nephron.

urine is manufactured is arranged around the mouth of the funnel and from this funnel the ureters pass the urine into the bladder for storage (Fig. 7.2).

The nephrons

The solid portion of the kidney consists of hundreds of minute twisted tubules each one of which starts as a little cup-shaped filter (Fig. 7.3). These tubules are called **nephrons** and they also resemble funnels and tubing (Fig. 7.4). Although these nephrons are only visible under the microscope they have very important work to carry out.

Each kidney has a large renal artery entering it. This artery becomes smaller and smaller until capillaries are formed. These capillaries form bunches in the mouth of the funnel part of each nephron. The funnel acts as a **filter** because it is made of porous

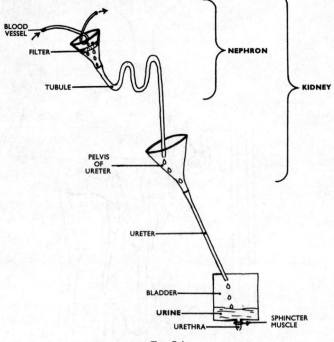

FIG. 7.4
Diagram illustrating formation of urine.

epithelium, and because of the pressure of the blood in the capillaries certain substances are forced through into the tubules. The filter, however, holds back the albumen in the plasma and also the blood cells. The fluid thus formed undergoes changes in composition as it flows along. This is the function of the specialized cells which form the walls of the tubules. They remove such substances as glucose and extra water and add things like toxins and drugs. Most of the drugs you give your patients are excreted by the kidneys and that is why they must be repeated regularly every four hours or three times a day as ordered.

The urine

The fluid which is formed in the nephrons is urine and consists of 96 per cent. of **water;** the remaining 4 per cent. is made up of **urea** (the waste product from protein foods), and **salts.**

In renal or kidney diseases, and also in a great many other conditions, examination of the urine is an important aid to diagnosis. This is why you are so often asked to save a specimen. What abnormalities do you find and why? Sugar is present in the urine of a diabetic patient because his blood contains too much. Albumin and blood are present in nephritis (inflammation of the nephrons) because the filter is damaged and no longer working. Bile is present in the urine of a patient who is jaundiced, because for some reason or other this substance which should pass through the intestines and be excreted in the faeces gets into the blood stream and the kidneys filter it out. There are, of course, other abnormal substances found in urine but many of these can only be discovered with the aid of a microscope in a laboratory.

THE URETERS

The ureters are very fine muscular tubes which leave the kidneys along with the renal veins. They carry the urine, or filtrate, down to the bladder and the veins carry the filtered blood into the vena cava.

THE BLADDER

When the bladder is empty it is a pelvic organ. It is a muscular bag lined with mucous membrane and lying in front of the other pelvic organs. As it fills up with urine it stretches and comes to lie in the abdominal cavity above the symphysis pubis where it can be palpated. It acts as a **reservoir** for the urine and will hold up to 300–450 ml quite comfortably. After that the individual has a feeling of fullness and there is a desire to empty the bladder. The act of passing the urine from the bladder is called **micturition.**

THE URETHRA

This is the passage from the bladder to the outside. It is short in a female, about 1½ inches (3 to 4 centimetres) in length, but much longer in a male when it is also part of the reproductive system. The exit from the bladder is guarded by a round sphincter muscle which must relax before micturition can take place.

You must have heard of **retention** of urine. Two possible causes of this are spasm of the sphincter muscle or pressure on the urethra which prevent micturition. In retention the bladder will go on stretching until it contains one to one and a half litres of urine and the patient becomes greatly distressed. You will learn in your nursing lectures various methods which can be used to persuade the tight muscle to relax.

Incontinence of urine is the result of a weak or paralysed sphincter muscle. This occurs in injuries and disease of the brain and spinal cord and in old age. The bladder now acts like a leaking bottle and urine dribbles out continually, making the nursing of such patients very difficult indeed.

Suppression of urine is different from retention and is a much more serious condition. In this case the kidneys are not manufacturing urine therefore the blood is not being filtered and its composition will be altered. The patient may die of a condition called uraemia where there is too much urea in the blood plasma.

Most kidney diseases have **oedema** as a symptom. Oedema is an increase of the amount of fluid in the tissues and in nephritis it is present because the diseased kidneys are not able to do their work properly. The oedema appears as a puffy swelling of the face and ankles.

The formation of **renal calculi** is another fairly common abnormality. Your patients call these 'stones', which is exactly what they are. They are formed in a sluggish urinary system by crystals of calcium salts. You can do a lot to help prevent their formation by moving your patient as often as possible and seeing that he drinks plenty of fluids.

Questions

(1) Several organs excrete waste products. Which one is wrong?
(a) skin (b) bowel (c) gallbladder (d) lungs (e) kidneys.
(2) The outer layer of the skin is called the ....
(3) The inner layer of the skin is called the ....
(4) The skin gets rid of waste products in the form of .... from the .... glands.
(5) The skin is waterproof because of the oily .... from the .... glands.

(6) The sensations conveyed from the skin to the brain are .... and ....
(7) The nerves of touch are contained in the: (a) dermis (b) epidermis.
(8) The skin regulates body temperature because the evaporation of .... uses up ....
(9) Which of the following causes an increase in sweating: (a) cold weather (b) excitement (c) a rising temperature?
(10) Vitamin .... is manufactured in the skin.
(11) The organs of the urinary system are: (a) ....
   (b) .... (c) ....
   (d) ....
(12) The kidneys lie at the .... of the .... cavity.
(13) The hollow part of the kidney is the ....
(14) The solid part of the kidney is made of ....
(15) The function of the kidney is to .... the blood.
(16) The substances removed from the blood by the kidney are: (a) .... (b) ....
   (c) ....
(17) Which of the following is the waste product excreted by the urine: (a) hormones (b) amino acids (c) urea (d) calcium?
(18) How much of the urine is water? (a) 80 per cent. (b) 96 per cent. (c) 50 per cent. (d) 67 per cent.?
(19) The ureters are made of ....
(20) The bladder lies in the ....
(21) The amount of urine which the bladder can comfortably hold is ....
(22) .... is the word which means the passing of urine from the bladder.
(23) This act is controlled by a .... muscle.
(24) If the bladder is full and no urine can be passed the patient suffers from .... of urine.
(25) If the bladder is empty and no urine is passed the patient suffers from .... of urine.
(26) In nephritis one of the following may be found in the urine: (a) glucose (b) bile (c) albumen (d) acetone.
(27) In diabetes one of the following may be found in the urine: (a) bile (b) acetone (c) blood (d) albumen (p.   ).
(28) Stones in the kidney are called ....

8. The Nervous System and the Special Senses

A good nurse must be observant. That is one of the first lessons you learn in nursing. What do you mean by being observant? To pay attention to what is going on around you, to keep your eyes open and see things? Yes, that is being observant; but I am sure your ward sister will tell you that it means much more than that. When you write your first night report about a very ill patient, it is not enough to describe what you see. Probably the patient lay still and slept all night long, but what about that hacking cough you heard? Did you smell his foul breath and feel his hot dry skin? Sister will consider these observations very important and from your report she will be able to sum up his condition, even before she sees him.

To give a full report you must therefore use all your special senses, with the possible exception of taste. Your eyes, your ears, your nose and your skin must all play their part in observation. As you read and learn more your brain will eventually, like Sister's, be able to make sense out of these observations. You will know that they are all signs of disease and you will learn what to do to help your patient.

You have gone through this learning stage before, only your mother took the place of Sister. Your infant eyes saw coal but it was your mother who told you that it was hot and dirty and that you must not touch it. You cannot remember when you first tasted an orange, heard a dog bark or smelt grilled bacon, but there must have been a first time. You know now what these things are and what they mean because you have a brain and that marvellous organ has stored away memories of all the impressions it has received from your sense organs.

Your life is one big collection of sights, sounds, tastes, smells, pain and of the feel of things; their texture and their temperature. Because you have a brain you can make use of these impressions; they form your intelligence. The lower animals have the same sense organs but not so much brain and that gives us a considerable advantage over them. They are guided almost entirely by their instincts or reflexes and have very little control or reasoning power. You too, in infancy, were guided by your instincts, you yelled when you were hungry, cried when you were hurt and emptied your bladder when it was full. Fortunately the majority of us learn some control because we are supplied with an elaborate and reliable nervous system.

The Nervous System

The nervous system is divided into two parts. One part receives impressions from the sense organs and sends out messages to the muscles of the skeleton: this is the central nervous system. The other part is the autonomic nervous system which controls functions of which you are not normally aware, for example, the movements of the internal organs and the heat regulating function of the skin.

THE CENTRAL NERVOUS SYSTEM

This consists of the **brain**, the **spinal cord** and the **nerves**. All three parts are made up of grey and white matter. Grey matter consists of cells which receive and send out messages. White matter consists of fibres each one of which is attached to a cell. The whole thing is like a complex telephone exchange with messages passing along the fibres, up and down the cord, to and from the brain.

A cell with its fibres is called a **neurone** (Fig. 8.1). There are two types of neurone:

Sensory neurones carrying messages in to the brain.

Motor neurons carrying messages away from the brain to the muscles.

The sensory neurones start as nerve endings in all the sensory organs and messages pass along their fibres to the brain. These are sensations of touch, pain and temperature from the skin, taste from the tongue, sight from the eyes, hearing from the ears and smell from the nose.

The motor neurones start as cells in the brain and their fibres end in the muscles. In this way the brain controls the movements of the body.

The brain
The brain consists of three parts:
   The *Cerebrum.*
   The *Brain Stem.*
   The *Cerebellum.*

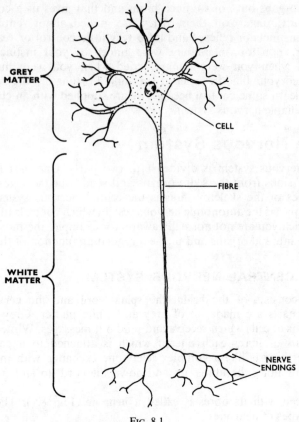

Fig. 8.1
A neurone.

The cerebrum is the largest part (Fig. 8.2). It consists of two hemispheres made up of grey matter on the outside and white matter on the inside. The hemispheres are joined by a bridge of fibres under which lie the **ventricles** of the brain. These are spaces which contain a clear fluid called **cerebrospinal fluid.** This cerebrospinal fluid comes from the blood capillaries in the ventricles and flows round the brain and spinal cord, nourishing them, carrying away their waste products and protecting them.

The surface of the cerebrum, the **cerebral cortex,** has a wrinkled appearance, like a walnut. All these folds increase the amount of grey matter and this is important because the cerebral cortex contains many specialized areas with important functions.

Extending from the top of each hemisphere in a broad strip down to the level of the ears is the **motor area.** The cells of this part control the movements of the skeletal muscles. Immediately behind this area is the **sensory area** which receives impulses from the skin. Below these are the areas for **taste, smell,** and **hearing.** Right at the back of the cerebrum are the areas for **sight.** In front, just behind the forehead, are the frontal lobes. These contain the cells which act as our filing cabinets storing away memories of things seen, heard, felt, tasted and smelt. This is the part with which we think, reason and remember. It is, in fact, our **intelligence** and **emotional** centre.

The brain stem. The nerve fibres which make up the white matter

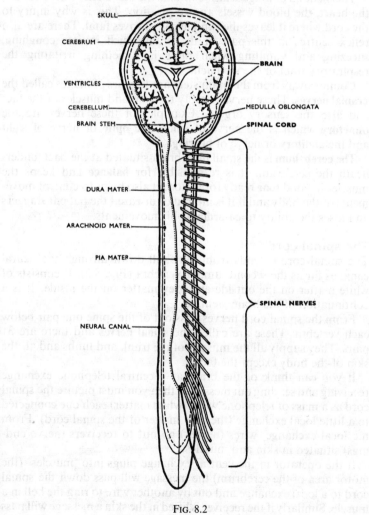

Fig. 8.2
Central nervous system.

of the cerebrum pass in towards the centre and downwards to form the brain stem.

This is like a stalk composed of fibres travelling up and down from the cerebrum to the spinal cord. The lower part of it is called the **medulla oblongata** and it is here that the fibres cross over so that the left hemisphere controls the right side of the body and vice versa. The next time you nurse a patient with a cerebral haemorrhage who has paralysis of the left side of his body you will remember that the injury must be to the right side of his brain.

The medulla oblongata also contains the nerve cells which control the heart, the blood vessels and respiration. This is why injury to the cord where it leaves the skull usually proves fatal. There are also reflex centres in this part of the brain which initiate coughing, sneezing and vomiting when there is something irritating the respiratory tract or the stomach.

Coming away from the brain stem are 12 pairs of nerves called the **cranial nerves.** These nerves supply the skin and muscles of the face and also the sensory organs. Examples of these nerves are the **olfactory** which is the nerve of **smell,** the **optic** or nerve of **sight** and the **auditory** or nerve of **hearing.**

The cerebellum is the small hind brain situated at the back underneath the cerebrum. It is responsible for **balance** and keeps the muscles in good **tone** ready for action. It also helps to control movements of the body and if it is injured or diseased the patient staggers and loses the ability to **co-ordinate** his movements.

The spinal cord

The spinal cord extends from the skull down through the neural canal as far as the second lumbar vertebra (Fig. 8.2). It consists of white matter on the outside and grey matter on the inside. It is a continuation of the brain stem.

From the spinal cord nerves pass out of the spine one pair below each vertebra. These are called the **spinal nerves** and there are 31 pairs. They supply all the muscles of the trunk and limbs and all the skin of the body except the face.

If you can think of the brain as a central telephone exchange receiving and sending out messages, then you must picture the spinal cord as a mass of telephone wires (white matter) each one connected to a little local exchange (the grey matter of the spinal cord). From the local exchange, wires (nerves) go out to receivers (nerve endings) situated in skin and muscles.

If the operator in the central exchange plugs into 'muscles' (the motor area of the cerebrum) the message will pass down the spinal cord to a local exchange and out by another wire to ring the bell in a muscle. Similarly if the receiver is lifted in the skin a message will pass up via the spinal cord to ring a bell in the brain. In this way the brain receives sensations from the skin and other sense organs and sends out messages to control the movement of muscles.

If your telephone has a dialling system you will know that you can put through a call without contacting the central exchange. Our nervous system has also a dialling system. It is usually called **reflex action** (Fig. 8.3). If the skin is in a hurry to get a message through to the muscle, as when you touch something hot, the message simply goes to the local exchange in the cord, out to the muscle, and the hand is withdrawn. This is rather like dialling 999 when there is a fire but fortunately with very much quicker results!

This comparison is interesting when we think of injuries to the nervous system. If the telephone wires are cut the telephone bell will not ring. If the nerves are cut no message of heat, cold, texture and pain will reach the brain and we say there is loss of sensation

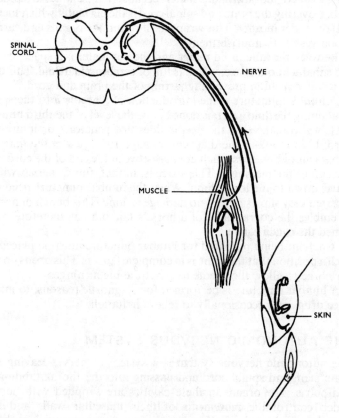

FIG. 8.3
Reflex action.

or anaesthesia of the skin. Similarly, no movement message can get out to muscle and the muscles supplied by that nerve are paralysed and will not contract. In injuries to the brain or spinal cord the central exchange is cut off but the local exchanges are still working. The operator has no control over what is going on. As the brain is our operator there is now no control over the movement of muscles, which perform purposeless movements and are said to be spastic. This spastic paralysis is seen in cerebral palsied children and in patients who have had injuries or diseases of the brain or spinal cord.

The meninges

The meninges are the protective coverings of the brain and spinal cord (Fig. 8.2). There are three of them and they have very descriptive names. The outer one lines the skull and neural canal and is called the **dura mater** or tough mother. Just underneath this is a thin membrane called the **arachnoid mater** because it resembles a spider's web. Covering the brain and spinal cord is the **pia mater** which means soft or tender mother. This wraps itself round the organs and carries blood vessels to nourish them.

Between the arachnoid mater and the pia mater is a space, called the subarachnoid space, which is full of **cerebrospinal fluid**. This acts as a water cushion preventing jarring of the brain and cord.

A lumbar puncture is the introduction of a needle into the space containing the fluid and it is done below the level of the third lumbar vertebra to make sure the needle does not penetrate or injure the cord. If you look at the diagram on page 107 you will see that the subarachnoid space is much greater between the end of the cord and the end of the meninges. This space is, in fact, full of nerves which travel down to the lower limb. When the lumbar puncture needle is inserted they slip aside and no damage is done. This bunch of nerves resembles the coarse hairs of a horse's tail and has therefore been named the **cauda equina**.

A patient being prepared for lumbar puncture must be placed in such a position that the spine is in complete flexion. This opens up the vertebrae to allow the needle to penetrate the meninges.

A lumbar puncture is performed for diagnostic reasons, to introduce drugs and occasionally to relieve irritation.

THE AUTONOMIC NERVOUS SYSTEM

The autonomic nervous system is a system of nerves leaving the brain stem and spinal cord and passing into the thorax, abdomen and pelvis. The organs in these cavities are supplied with nerves which control the **movements** of their muscular walls and the **secretions** of their glands. These are movements over which we have no control. This system also sends branches to the skin to supply the little muscular blood vessels, the muscles which make your hair stand on end, and the sweat glands.

When we are apprehensive, afraid, or making a great effort our hearts must beat faster to supply the voluntary muscles and the brain with blood. At the same time we experience 'butterflies in our tummies' and the thought of food makes us feel sick. This is Nature's way of sharing out the vital oxygen and seeing that the organs concerned with action are supplied while those concerned with food are deprived.

Peace and absence from worry are the best aids to digestion because the autonomic system will then be able to see that the largest share of oxygen goes to the digestive organs. A patient who is worried and upset will not be able to digest his food properly so you must see what you can do to give him peace of mind.

The nerves which quicken the heart beat and slow digestion are called the **sympathetic** nerves. Those which slow the heart beat and quicken the digestion are called the **para-sympathetic** nerves.

The Organs of Special Sense

In preceding chapters we have learned about the skin, the nose, and the tongue, so now we must consider the two remaining sense organs, the eye and the ear.

THE EYE

This is the organ of sight and is concerned with receiving rays of light reflected from the objects around us. It passes the impressions received back via the optic nerve to the areas of the cerebrum which tell us what we see.

The eyes are almost spherical and they lie in the cone-shaped bony orbit protected by the eyelids in front and by a pad of fat at the back. They are attached to the bone by small muscles which explains why they can be moved about in so many directions. Weakness of one of these muscles will mean that the eye squints.

Dissecting an eye is rather like cutting through an onion only it is not solid the whole way through. It has three coats, but the centre is full of a jelly-like fluid (Fig. 8.4).

The outer coat

This is the protective coat of the eye and it has the eye muscles attached to it. It consists of two parts. The larger part, at the back, is white and opaque and is called the **sclera**. In front is the clear transparent **cornea**, like a window, through which the rays of light pass. The cornea is very sensitive and the slightest spot of dust touching it

112 Anatomy and Physiology as Applied to Nursing

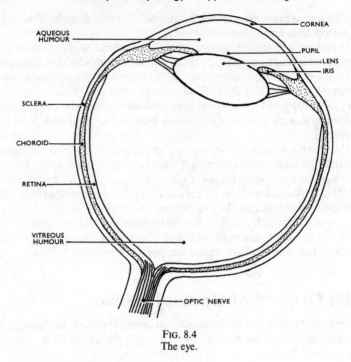

Fig. 8.4
The eye.

will bring down the eyelid like a shutter. This blinking helps to keep the window clean.

The middle coat
This is the coloured part of the eye and it contains the blood vessels. It is dark in colour like the inside of a camera. This prevents the rays of light which have entered from being reflected out again. It is called the **choroid** and attached to it in front is the coloured part of the eye called the **iris**. This is like a circular curtain with a hole in the centre. This hole is called the **pupil** and as the iris is muscular the size of the pupil can be altered. It is like the shutter of a camera controlling the amount of light which enters. If it is a sunny day you screw up your eyes and shade them but at the same time the autonomic nervous system is constricting the pupil and keeping out some of the light. On the other hand, at twilight, before you put the lights on, the pupils will be dilated allowing more light to enter.

Certain drugs have an effect on the size of the pupil; morphia constricts it and atropine dilates it. Homatropine drops are occasionally instilled into an eye to allow the doctor to get a better view of the inside with an ophthalmoscope. After this has been done another drug, eserine, must be instilled to return the pupil to normal. In the nursing of patients with head injuries accurate observations must be made of the pupils as their reactions are of great value in diagnosis.

Situated just behind the pupil is the **lens**. This is like a crystal allowing the rays of light to enter and at the same time bending them, so that they can be focused on the most sensitive part of the nervous coat. The shape of the lens can be altered by a small muscle inside the eye and this allows us to see distant objects as clearly as near ones. This is called accommodation. Sometimes the lens becomes hardened in the centre and is now opaque instead of clear. This white spot on the lens is called a cataract and is a common cause of blindness in old age.

The inner coat

This is the nervous coat or **retina**. It resembles the film of a camera because it is here that an upside down image of the object seen is formed. Fortunately everything we look at is turned right way up by the brain which develops the film and prints it.

The retina contains the nerve endings which are stimulated by the rays of light. The impulses then pass back in the **optic nerves** to the posterior lobes of the cerebrum. When we look directly at an object it is much clearer than its surroundings. Try looking at the wall above the clock; can you tell the time? No, because the rays of light from the wall are being focused on the most sensitive part of your retina and not the rays from the clock. This sensitive area is right at the back of the eyeball just beside the spot where the nerve leaves. It appears as a **yellow spot** and has more numerous nerve endings than the rest of the retina.

Injury to any part of the pathway from the eye to the posterior lobes of the cerebrum may result in partial or complete blindness.

THE STRUCTURES WHICH PROTECT THE EYE

The eye is a very delicate organ and we have already seen that it lies in the body orbit with a protective **pad of fat** behind it. In old age and long and serious illness this pad of fat diminishes and gives the patient a hollow-eyed appearance. On the other hand, in exophthalmic goitre this pad of fat enlarges and pushes the eyeball forwards.

The other structures which help to protect the eye are the eyelids, the eyebrows and the tears. The **eyelids** are like shutters; they are muscular, covered with skin and can be opened and closed at will. The eyelashes are small hairs growing from the edge. When any object approaches the eye suddenly the lids automatically shut and the eyelashes help to sweep it away, preventing injury to the eye.

Covering the front of the eyeball and lining the eyelids is a thin layer of mucous membrane called the **conjunctiva**. Any insect or speck of dust which has not been blinked away by the eyelashes or eyelids will stick to the conjunctiva and can be removed without damaging the eye itself.

The **eyebrows** help to cut down the amount of light entering the eye. This is why you frown when the light is too bright. They also prevent sweat from the forehead running down into the eye.

A foreign body in your eye makes you weep, not because it is so painful but rather as an attempt to irrigate and wash away the irritating object. Situated above each eye is a little gland, the **lachrymal gland,** which is continually producing a salty fluid. These **tears** flow over the eye, washing it, and are evaporated or pass down the lachrymal duct to be evaporated in the nose. It is only when the glands produce additional fluid, as when we are emotionally upset, that we are aware of tears. Tears are like the water running down the fishmonger's window continually keeping it damp and clean.

Sight and seeing

Rays of light from objects hit the transparent cornea and are then bent as they pass through the watery fluid **(aqueous humour)** lying between it and the lens. They are further bent by the lens and pass through the jelly-like centre **(vitreous humour)** to meet at the yellow spot on the retina. At this point the nerve endings are stimulated and impulses pass by the **optic nerves** to the posterior lobe of the cerebrum where these rays are interpreted as sight. If the rays are bent too far or not enough, the brain does not receive a clear picture of the object and glasses must be worn to correct this short or long sight.

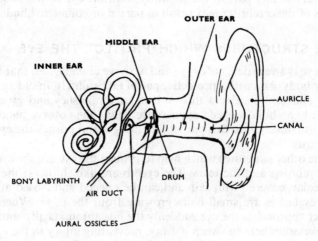

Fig. 8.5
The ear.

THE EAR

This is the organ of hearing and is constructed in such a way that waves of sound passing through the atmosphere are caught up and transmitted to the hearing centre in the cerebrum. It also has a part to play in maintaining balance.

The visible part of the ear, the auricle or trumpet, is made of cartilage and is of more use to the lower animals because they can move them towards the direction from which the sound is coming. The more important part of the ear lies within the temporal bone; it consists of three parts:
> The outer ear.
> The middle ear.
> The inner ear.

The outer ear
This consists of a short twisted **canal** leading from the auricle to the ear **drum** which separates it from the middle ear (Fig. 8.5). It is lined with skin which contains hairs and little glands secreting wax. This helps to guard against the entrance of insects and dust. This is also the reason why the canal is twisted. If you watch the doctor irrigating an ear you will see that he first of all straightens out the canal by pulling the auricle of the ear backwards and upwards otherwise the fluid will not reach the end of the canal.

The middle ear
This is rather like a little room carved out in the bone (Fig. 8.5). The outer wall contains one large window and the opposite wall a very small one. These windows are completely covered over with membrane, the larger one is called the **drum** and the smaller one the **fenestra ovale** (the oval window). Extending right across this chamber from the drum to the oval window are three little bones called the **aural ossicles.**

Entering the chamber is an **air duct** which brings air from the nose into the middle ear. This keeps the atmospheric pressure on either side of the drum equal. When there is any sudden alteration in pressure as happens for example in flying, when the aeroplane is coming down, the increased pressure on the outside of the drum makes you feel deaf. Unfortunately this duct can just as easily carry infection and occasionally middle ear infection or otitis media complicates a cold in the nose or a sore throat.

The inner ear
On the far side of the small oval window is a complicated arrangement of passages cut out of the bone and called the **bony labyrinth** (Fig. 8.5). This is full of fluid in which floats a **membranous**

labyrinth. Inside the membranous labyrinth is more fluid in which lie the nerve endings which are the beginning of the nerve of hearing and balance.

Hearing and balance

Waves of sound pass through the canal, hit the drum which then vibrates and moves the three aural ossicles. The last little bone hits the oval window and this vibration starts movement of the fluid in the bony labyrinth. This in turn moves the membranous labyrinth and stimulates the nerve endings which are floating in the fluid. The impulses received by the nerve endings pass back to the cells in the hearing area of the cerebrum.

Some of these nerve endings are stimulated not by waves of sound but by the pressure of the fluid changing with different positions of the head in space. This is the way in which the ear helps to control balance. If you keep changing your position quickly as, for example, during an old-fashioned waltz, you will be giddy because of overstimulation of this part of the ear.

Questions

(1) The two parts of the nervous system are: (a) ....
   (b) ....
(2) Nerve cells are .... in colour.
(3) Nerve fibres are .... in colour.
(4) A nerve cell with its fibre is called a ....
(5) A neurone which carries messages from the brain is: (a) sensory
   (b) motor (c) autonomic.
(6) There are seven different sensations. These are:
   1. .... 2. ....
   3. .... 4. ....
   5. .... 6. ....
   7. ....
(7) The three parts of the brain are: (a) ....
   (b) .... (c) ....
(8) The surface of the cerebrum is the ....
   ....
(9) The area of the cerebrum responsible for sight is: (a) at the side (b) on the top (c) at the front (d) at the back.

The Nervous System and the Special Senses

(10) The part of the brain which controls balance is the: (a) cerebrum (b) cerebellum (c) brain stem.
(11) The lowest part of the brain stem is called the ....
(12) The centre which controls respiration is in the ....
(13) The grey matter is on the .... of the spinal cord.
(14) The spinal cord carries messages from the brain to .... and from the .... to the brain.
(15) The spinal cord can also carry impulses from skin to .... without going to the brain. This is called the ....
(16) The spinal cord finishes at the: (a) 1st sacral vertebra (b) 2nd dorsal vertebra (c) 2nd lumbar vertebra (d) 1st dorsal vertebra.
(17) The coverings of the brain and the spinal cord are the: .... There are three and these are the
(a) ....
(b) ....
(c) ....
(18) Cerebrospinal fluid lies in the .... space.
(19) The autonomic nervous system controls the .... of glands and the .... of internal organs.
(20) There are two parts to the autonomic nervous system, the .... and the ....
(21) The heart beat is quickened by the .... nerves.
(22) The cone-shaped cavities in the skull which contain the eyes are called the ....
(23) The cavities are lined with ....
(24) The transparent part of the outer coat of the eye is the: (a) choroid (b) cornea (c) sclera (d) lens.
(25) The iris is the .... part of the eye.
(26) The pupil dilates in: (a) bright light (b) dim light (c) morphine poisoning.
(27) Accommodation depends on the shape of the ....
(28) The nervous coat of the eye is the ....
(29) The most sensitive part of the nervous coat is the ....
(30) The covering of the eye is the: (a) sclera (b) conjunctiva (c) retina (d) choroid.
(31) The .... glands secrete tears.
(32) The ear has two functions, hearing and ....

(33) The drum separates the outer from the .... ear.
(34) The small bones in the middle ear are the ....
    ....
(35) The nerve endings of the nerve of hearing lie in the membranous
    ....
(36) The nerve ending for hearing is stimulated by ....
    of sound.
(37) The nerve endings for balance are stimulated by alterations in
    fluid ....

9. The Endocrine or Ductless Glands

Have you ever wondered why some of your friends are always doing something in their off-duty whilst others go to bed or are content to sit around gossiping? What makes you want to stay at home and finish your embroidery when your friends are all urging you to go out cycling or to watch a football match? Why does Nurse Smith never feel the cold even in the middle of winter? Why are you too frightened to open your mouth in class while Nurse Black is for ever interrupting the tutor?

What makes us all different? Inheritance, background, upbringing and endocrine glands all play a part in making us what we are.

These endocrine glands are very complex and many of their functions are not completely understood. It is by studying what happens when they become diseased that their functions are being discovered.

Unlike the sweat glands and the digestive glands these endocrine glands have no duct or channel carrying their secretions to the parts where they are required. Instead, they put their secretions directly into the blood which is flowing through them. This is why they are called ductless glands and their secretions or **hormones** are called internal secretions.

The following are some of the more important endocrine glands:
The *Pituitary Gland*.
The *Thyroid Gland*.
The *Parathyroid Glands*.
The *Adrenal Glands*.
The *Ovaries* and the *Testes*.

The pituitary gland
This is a small gland about the size and shape of a cherry. It lies in the skull at the base of the brain. Although smaller than the other glands, it seems to be more active, produces many hormones and

has some control over the activities of the other glands. It has two lobes, an anterior and posterior lobe.

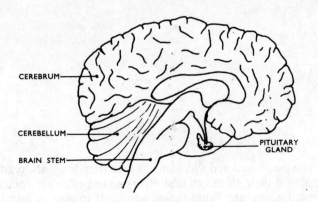

Fig. 9.1
The pituitary gland.

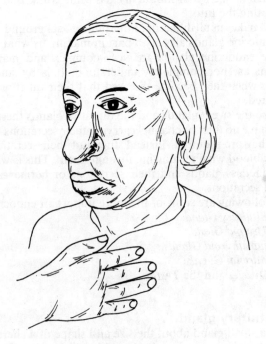

Fig. 9.2
Acromegaly.

The **anterior lobe** produces a hormone which affects **growth**. A child born with a deficiency does not grow normally and remains a **dwarf**. A child born with too much of this hormone grows too tall and becomes a **giant**. In adults, oversecretion produces thickening of the hands, the feet and skull. This condition is called **acromegaly** (Fig. 9.2). This anterior lobe also produces other hormones which control the development of the **sexual organs** and stimulate the **thyroid** and **adrenal** glands. Under the influence of this lobe of the pituitary gland the breasts will produce milk after the birth of an infant.

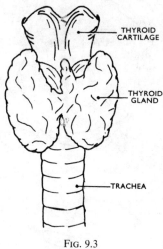

FIG. 9.3
Thyroid gland.

The posterior lobe of this gland produces three hormones. One called the **antidiuretic hormone** controls the amount of water excreted by the kidneys. The others act on the muscular walls of internal organs. These hormones act on the small arteries, contract their walls and **raise the blood pressure,** contract the wall of the bowel and **increase peristalsis** and contract the walls of the uterus during childbirth.

An extract of the posterior lobe of the pituitary gland called **Pituitrin** is given medically to raise the blood pressure in shock, to overcome an obstruction in the bowel and to assist a mother to give birth to her child.

The thyroid gland
This gland consists of two lobes which lie one on each side of the larynx. It is normally invisible, but, if enlarged, forms an obvious swelling at the base of the neck. It produces a hormone called

thyroxine. Iodine is necessary in the diet for the proper formation of thyroxine. It is the only tissue to require this mineral salt. Thyroxine controls the rate of metabolism (p. 89). It is necessary for normal mental and physical development and for healthy hair and skin. An underactive thyroid produces a lazy individual, slow in his movements and thoughts; a person who is fat and puffy with coarse dry skin and hair. He is said to suffer from **myxoedema** (Fig. 9.4).

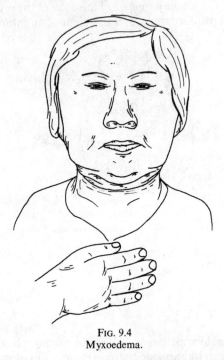

FIG. 9.4
Myxoedema.

If an infant should be born with an inefficient thyroid it is called a **cretin** (Fig. 9.5). This child also has a dry, coarse skin, straight black hair, a prominent abdomen, is mentally deficient and has a tongue which appears to be too big for his mouth. Fortunately these people can be treated by giving them thyroid extract for the rest of their lives.

An overactive thyroid gland produces an overactive individual. They burn up their fuel foods at a terrific rate. They have good appetites but are always thin. Their eyes protrude and they are excitable and nervous and have a very rapid pulse rate. They are always warm because they make so much heat in their own bodies. These people can also be helped but sometimes only by operative removal of part of the gland. This condition is called **hyperthyroidism** or **exophthalmic goitre** (Fig. 9.6).

Fig. 9.5
A cretin.

The parathyroid glands
There are four little parathyroid glands situated on the back of the thyroid. Their hormone regulates the use the body makes of **calcium**. If the glands are overactive the amount of calcium in the bones is reduced and the bones are soft and painful. Under-secretion of this hormone causes muscular spasms in the hands and feet. This is tetany and is due to insufficient calcium in the blood.

The suprarenal or adrenal glands
These glands are situated above each kidney but have nothing to do with their function. They consist of two distinct parts, the cortex on the outside and the medulla in the centre.

The hormones from the cortex are very complicated. They are called **steroids**. One of the steroids controls the amount of **salt** in the body. Undersecretion upsets the balance of salt, there is too little in the blood and the patient suffers from a sometimes fatal disease

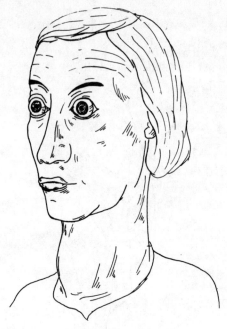

Fig. 9.6
Hyperthyroidism.

called Addison's disease. Oversecretion, as the result of a tumour, produces various **sexual changes** depending on the age and sex of the individual. A woman may become masculine and a man feminine; a child will reach puberty at a very early age.

Cortisone is another of the steroids from the adrenal cortex. Its function is to help the body to resist stress and to prevent permanent damage to various tissues resulting from injury, strain and emotion. It is used medically in the treatment of a wide variety of disorders including skin conditions and arthritis.

The medulla of the adrenal glands produces a hormone called **adrenaline.** This substance is secreted when you are upset or afraid. It is called the stress hormone and works in a similar way to the sympathetic nervous system. Adrenaline is shot into your blood prior to an examination or an important interview, the blood pressure shoots up, the pulse is rapid, more oxygen is taken into the lungs and vital glucose is liberated from the store in the liver. The very last thing you want is a big meal and you may even feel sick. There is no need to describe the symptoms as everyone is familiar with them. It is comforting to think that this is Nature's way of helping you to face up to the situation and that your brain is getting lots of vital glucose and oxygen.

The ovaries
These are the female sex glands and will be described in the next chapter. They secrete the hormones which bring about the changes occurring in a girl at puberty and maintain the normal menstrual cycle in a woman during her child bearing years.

The testes
These are the male sex glands. The hormone is responsible for the changes occurring in a boy at puberty and for maintaining the normal sexual urge in an adult male.

Other endocrine glands
There are other endocrine glands such as the **pineal** in the brain and the **thymus** in the thorax but their functions are not yet completely understood.

The **pancreas,** although a gland with a duct, is also an endocrine gland. There are little areas of special cells scattered throughout the pancreas called the **islets of Langerhans.** These cells produce a hormone called **insulin** which allows the body to store glucose as glycogen. If it is absent or deficient the patient suffers from diabetes and most of the glucose is lost by being excreted by the kidneys.

Questions
(1) The secretions of endocrine glands are ....
(2) These secretions go directly into the ....
....
(3) Which of the following is not an endocrine gland: (a) the thyroid (b) the parotid (c) the pituitary (d) the adrenal?
(4) The pituitary gland lies: (a) in the head (b) in the thorax (c) on top of the brain (d) at the base of the brain.
(5) An adult patient suffering from .... has oversecretion of the pituitary gland.
(6) A cretin is deficient in the hormone from the .... gland.
(7) An adult with deficiency of the thyroid hormone suffers from: (a) acromegaly (b) hyperthyroidism (c) diabetes (d) myxoedema.
(8) An adult suffering from exophthalmic goitre has: (a) a rapid pulse (b) prominent eyes (c) a poor appetite (d) a nervous disposition. Which one is wrong?
(9) A healthy thyroid gland requires .... in the diet (Chap. 6, p. 00).
(10) The parathyroid glands produce a hormone which controls the metabolism of ....
(11) The adrenal glands lie above the ....

(12) The hormone used to treat skin conditions is ....
(13) The hormone which converts glycogen into glucose is ......
....
(14) The hormone which converts glucose into glycogen is ......
....
(15) The islets of Langerhans are part of the ....
and they secrete ....

10. The Reproductive Systems

The simplest of living things, for example, yeasts, moulds, and bacteria reproduce themselves by throwing out buds or simply dividing in two. The last method means that the parent disappears leaving behind two new organisms. Can you imagine the utter chaos if this happened in the higher animals? Every baby would have an identical twin and no parents.

In all the higher animals there is some form of sexual reproduction, each new creature having two parents, male and female. The essential feature of this method of reproduction is that the female produces the ovum or egg cell and the male the spermatozoon. These two cells must unite to form a new individual. In the human animal the union takes place within the mother's body and the fertilized egg cell is retained there until it has grown into a new individual capable of a separate existence.

The Female Reproductive System

As the female reproductive system has a very different function from the male it is more elaborate and takes up more room. This is one of the reasons why the female pelvis is bigger than the male. This system is constructed in such a way that it can produce hundreds of ova (one every 28 days) any one of which may be fertilized. The fertilized ovum divides again and again until it takes on the form of its parents when it is called a foetus. The foetus is retained for nine months and during this time it is nourished by the mother who also gets rid of its waste products. At the end of nine months it is ready to lead a separate existence and is expelled from the uterus.

The organs of the female reproductive system are:
The *Ovaries*.
The *Uterus*.
The *Vagina*.
The *Mammary Glands*.

The ovaries

These are two almond-shaped glands situated in the pelvic cavity one on each side of the uterus (Fig. 10.1). They do not begin to function until the girl reaches puberty, between 12 and 14 years of age. Every ovary is full of tiny cavities which contain fluid and an immature **ovum**. At puberty and every 28 days from then until the menopause (about 50 years of age) one little cavity comes to the surface, bursts and expels its ovum which is now mature. This ovum finds its way into the uterus where it may or may not be fertilized.

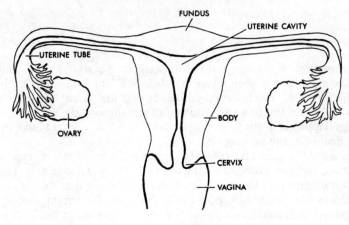

FIG. 10.1
Female reproductive system. Internal organs.

The ovary also produces the **female sex hormones, oestrin** and **progesterone,** which are responsible for the physical and mental changes in the adolescent girl and also for the preparation of the uterus and breasts for pregnancy.

The uterus

This is a hollow organ, about 4 inches (10 centimetres) long, situated between the bladder and the rectum. Projecting out from it on each side are two muscular tubes called the **uterine tubes** (Fig. 10.1). These end in an expanded part rather like the sucking end of a vacuum cleaner and this is attached by a ligament to the corresponding ovary. When the ovum escapes from the ovary it is 'sucked' up through the tube into the uterus.

The uterus has a very thick muscular wall which thins out as it expands during pregnancy. The non-pregnant uterus is pear-shaped. The broad portion lies uppermost between the Fallopian tubes and is called the **fundus.** The main part is called the **body** and this narrows at the bottom to form the **cervix** or neck which projects down into the vagina (Fig. 10.1).

The uterus is lined with a very special tissue called the **endometrium** and covered by a double fold of the **peritoneum** which is called the **broad ligament** because it helps to hold the uterus in place.

The menstrual cycle

From puberty to the menopause a series of changes take place in the uterus every 28 days. This is called the **menstrual cycle** and these changes are concerned with reproduction. Once every month a mature ovum is expelled from an ovary, finds its way into the uterine tube and is pushed along to the uterus. At the same time a hormone, progesterone, is produced by the ovary which thickens the endometrium in preparation for the ovum being fertilized by a spermatozoon (p.   ). Should fertilization occur the ovum burrows into the thickened uterine lining. There it starts to grow and forms the embryo which in turn becomes the foetus or unborn child. This hormone from the ovary stops the menstrual cycle for the nine months of the pregnancy.

If fertilization of the ovum does not occur, the thickened endometrium falls away, breaks up and is expelled. This is called the menstrual period or flow and it lasts for about five days. During the nine days following the end of the menstrual flow another hormone, oestrin, repairs the endometrium until the next ovum is ripe. This ripe ovum is set free approximately 14 days from the onset of the menstrual period. The next 14 days are concerned with the thickening of the endometrium for the fertilized ovum—rather like a bird lining its nest before laying eggs.

The vagina

The vagina is the passage leading from the uterus to the outside; it is a flattened muscular tube lined with mucous membrane. The opening of the vagina lies between the urethra and the rectum and is protected by two folds of skin on each side. These folds are the labia of the **external genitals** or **vulva** (Fig. 10.2).

The mammary glands

The mammary glands or breasts are present in both sexes but only develop in the female sex at puberty and do not function until the end of pregnancy.

Each gland consists of a number of tubules which open on to the surface by ducts at the nipple. The tubules are called **lactiferous tubules** because they secrete **milk** two or three days after the birth of the child. This milk is the ideal food for the baby as it contains all the essential food factors, is the correct temperature and is free from infection.

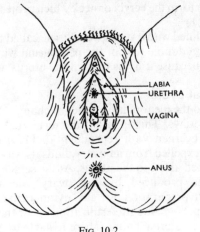

Fig. 10.2
External genitals.

The Male Reproductive System

The male reproductive system is responsible for the formation of the **spermatozoon** or male germ cell and for projecting it into the female vagina. The spermatozoa are small cells with long tails. This tail gives the cell its mobility and drives it up through the uterus into the uterine tubes where it may fuse with an ovum. Enormous numbers of these spermatozoa are liberated at a time.

The organs of the male reproductive system are:
The *Testes.*
The *Vasa Deferentia.*
The *Seminal Vesicles.*
The *Prostate Gland*
The *Penis.*

The testes

These are the glands which produce the **spermatozoa** and the male sex hormone, **testosterone**. The hormone produces the characteristic changes in the boy at puberty and also activate the testes to produce spermatozoa.

The two testes lie outside of the body in the **scrotum** but in the foetus they develop in the abdomen and only descend into the scrotum about one month before the baby boy is born. These glands are made of a number of lobes inside which are the tubules which produce the germ cells. These tubules are connected to the vasa deferentia by tiny twisted tubes (Fig. 10.3).

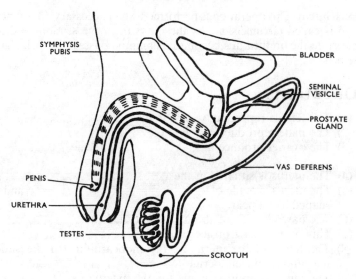

Fig. 10.3
Male reproductive system.

The vasa deferentia
These are hollow tubes which pass up from the scrotum into the pelvis and carry the spermatozoa into the seminal vesicles.

The seminal vesicles
The seminal vesicles are two little sacs lying at the base of the bladder; they secrete a fluid which is added to the spermatozoa to form **semen**.

The prostate gland
This gland surrounds the urethra just as it leaves the bladder. It puts its secretion into the semen which passes through this gland, in two little ducts to enter the urethra.

The prostate gland sometimes becomes enlarged in old age and because of its position obstructs the urethra and causes acute retention of urine.

The penis
The penis like the testes is an external organ; it lies in front of the scrotum.

The **urethra** runs through the penis which consists of three columns of specialized tissue with a rich blood supply. It ends in a bulbous structure called the **glans penis** which is protected by a fold of skin called the prepuce. This prepuce should be loose and if tight must be cut. This operation is performed when necessary, in infancy, and is called **circumcision**. As the ducts from the seminal vesicles pass into the urethra this becomes a common passage for urine and semen.

Questions

(1) The female reproductive cell is the ....
(2) The male reproductive cell is the ....
(3) The ovaries produce the .... and the ....
   ....
(4) The uterus is situated in the ....
(5) The uterus is made of .... and is shaped like a pear.
(6) The narrow neck of the uterus is called the ....
(7) This narrow part projects into the ....
(8) The lining of the uterus is: (a) endocardium (b) the endometrium (c) the endocrine.
(9) The vulva is another name for the external ....
(10) Milk is secreted after the birth of a child by the .... glands.
(11) This is controlled by the .... gland. (Chap. 9)
(12) The glands which produce the spermatozoa are the ....
(13) These glands also produce the ....
   ....
(14) The testes lie in the ....
(15) The vasa deferentia carry the spermatozoa into the ....
(16) Retention of urine may be caused by an enlarged ....
(17) The male urethra is in the ....

11. Posture—Nurse and Patient

The purpose of this chapter is to describe how, even with a very elementary knowledge of the structure of the body, much can be done to avoid pain and discomfort. Although the days of 'mopping the fevered brow' may be gone it is inconceivable that we will ever be able to dispense with pillows. Nor is it likely that mechanical hoists will ever completely take the place of the nurse's arms! A nurse must have a knowledge of posture. She must carry herself well, be able to lift heavy objects without injuring herself and, above all, she must be able to make her patient comfortable.

Good posture does not necessarily mean sailing up and down the ward with your head up and your shoulders back. It means balancing your body in such a way that all the bones, joints, ligaments and muscles are performing their tasks smoothly without one part taking more of the strain than the others. This does not only apply to standing and walking. The body should be well balanced when you are sitting, lying, lifting, carrying, bending, pushing or pulling.

The bones with their movable joints and muscle attachments not only protect the internal organs but provide a framework capable of all the intricate movements necessary for daily living. One of your first big accomplishments in life was to make that framework hold you upright. I am sure you must have watched a baby struggle to get his balance and take his first steps. From there he goes on learning more and more complex movements and the once difficult feat of balance is taken for granted.

Balance is the basis of the upright position and good posture depends on the ability to balance the body properly when standing, sitting or walking. The laws which govern body balance are the same as those which apply to pitching a tent. If you want your tent to stand upright the guy ropes must exert an equal pull on the tent pole and the further away the pegs are from the pole the more likely it is to stand (Figs. 11.1 and 11.2). The centre of gravity in the body is a line drawn from the top of the head which passes through the

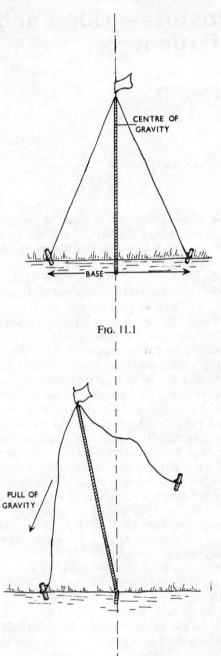

Fig. 11.1

Fig. 11.2

centre of the trunk and reaches the ground between the feet which form the base (Fig. 11.3). If the pull of muscles is not equal on either side or at the front and back, the body will bend over (Fig. 11.4). Sometimes with the spine held straight it is still difficult to balance as, for example, when you are standing in a moving bus.

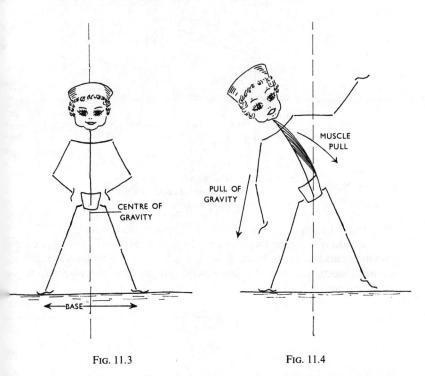

Fig. 11.3   Fig. 11.4

You will find that, without thinking, you place your feet wider apart to keep yourself from falling over when the bus stops.

Many times in the course of a day you bend over, but fortunately you rarely lose your balance. This is because the muscles on the side opposite to that to which you are bending go into tight contraction; the ligaments and discs of the spine take the strain and hold you back. This is all very well until you start lifting and carrying heavy objects in this position, then the muscles are forced into violent contraction and the ligaments and discs receive sudden jerks which may tear or displace them.

If your back aches after making the beds it is probably because your posture has been poor; you have been bending your spine in order to lift your patient and the muscles of your back have been strongly contracted to keep you from losing your balance (Fig. 11.5).

INCORRECT
Fig. 11.5
Posture when bedmaking
and lifting.

You should always get close to the bed when you are about to lift. You should place your feet moderately wide apart, bend your knees and hips and keep your back straight (Fig. 11.6). Your body is now well balanced, you will tire less easily and are less likely to strain or injure your back.

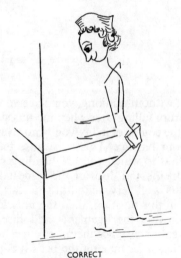

CORRECT
Fig. 11.6
Posture when bedmaking
and lifting.

As well as developing a habit of good posture in yourself you must observe similar rules for your patient. It is not a question of balance when lying in bed or sitting in a chair but of positioning the patient's body in such a way that no part takes unnecessary strain. Avoid having the patient's spine twisted to either side and remember to support the natural curves. The legs should be in the neutral position with the patellae facing directly upwards, the knees slightly flexed and the feet at right angles.

Patients who are active must be encouraged to change their position frequently, spending some part of the day in each of the positions which their condition allows them to adopt. This keeps the joints mobile, the muscles in good condition and helps the circulation of the blood. The helpless patient must, however, be moved in bed by the nurse who must observe his position and remember that all the bony prominences must be relieved of pressure.

We shall consider some of the positions used in nursing, some ways of maintaining good posture and the prevention of fatigue and discomfort. To achieve this it is necessary to have a good mattress which does not sag. This may be supported on a firm base or on fracture boards. In addition, a variety of pillows, a foot-board, a bed cradle and sandbags may be required.

The Supine Position

The supine position is lying on the back. In most cases the mattress, with a small pillow under the head, will give all the support which is necessary (Figs. 11.7 and 11.8). Some patients, however, may require a small pillow in the lumbar curve and each leg supported on individual pillows placed lengthwise (Fig. 11.9). These pillows keep the knees in slight flexion and prevent the heels from rubbing on the mattress. Leg pillows are not used unless absolutely essential as they restrict movement and if hard and bulky interfere with the circulation and may cause venous thrombosis. A board may be

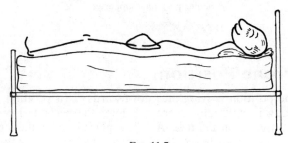

Fig. 11.7
Supine position—Unsupported. Good

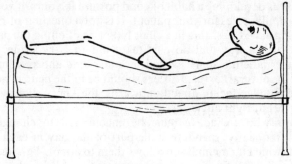

Fig. 11.8
Supine position—unsupported.
Bad

necessary to keep the feet at right angles because, if unsupported, they tend to drop. This is due partly to the pull of gravity and partly to the weight of the bedclothes. If the legs tend to roll outwards, sandbags may be placed lengthwise along the outside.

Fig. 11.9
Supine position—supported.

The Prone Position

The prone position is lying face downwards. This position forces the feet down, contractures occur behind the heels and pressure sores may appear on the toes. A pillow placed under the legs will keep the knees slightly flexed, the feet at right angles and at the same time take the pressure off the toes (Fig. 11.10). If there is any

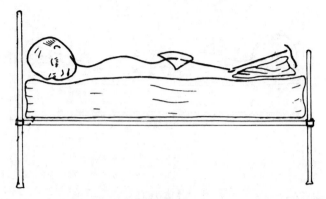

Fig. 11.10
Prone position.

possibility of flexion contractures occurring at the knees, the feet should be positioned over the end of the mattress (Fig. 11.11). A pillow may be placed under the head, which must be turned to one side, and another under the chest. If there is any danger of pressure sores developing, another pillow is placed across the bed under the pelvis to take the weight off the iliac spines (Fig. 11.12). This position is not popular with adult patients. You must make quite sure that the patient's breathing is unimpeded and that he can see what is going on around him.

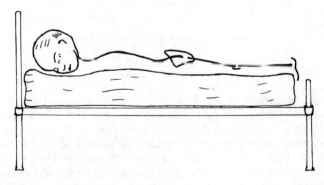

Fig. 11.11
Prone position.

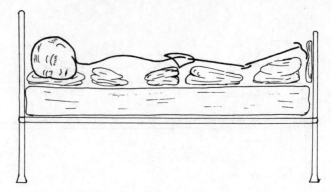

Fig. 11.12
Prone position—supported.

The Lateral Position

The lateral position is lying on the side. A weak helpless patient finds it difficult to maintain unless properly supported. A pillow

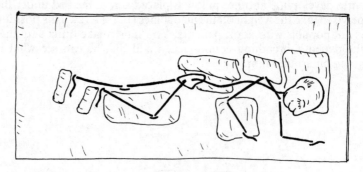

Fig. 11.13
Lateral position.

should be placed in the loin, relieving the ribs and pelvis from pressure. If this is brought round to the back and held secure by a sandbag, it will support the patient and keep him from rolling over. The uppermost leg should be placed on pillows so that its weight does not pull on the hip joint and twist the spine. This leg should not lie directly on top of the other but should be flexed forwards. If necessary, the feet can be kept at right angles by using sandbags (Fig. 11.13).

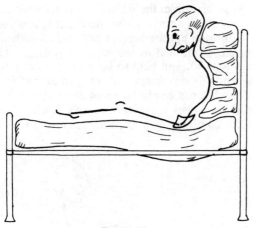

Fig. 11.14
Upright position—in bed.
Incorrect

The Upright Position

In this position the patient is sitting erect. It is a difficult position to maintain in bed particularly if the patient is heavy and tends to slip down. If incorrectly supported, posture can be very poor; weight is taken on the sacrum instead of on the ischial tuberosities, the spine is flexed and the chest is depressed (Fig. 11.14). To achieve

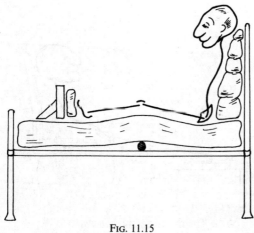

Fig. 11.15
Upright position—in bed.
Correct

the correct position and make the patient comfortable, the pillows should be arranged in such a way that the lumbar and cervical curves are supported but the head is not pushed forwards. A footboard fixed across the bed in the required position will not only prevent drop foot but will help to keep the patient upright. If the patient still tends to slip down and no cardiac bed is available, a piece of broomstick can be placed across the bed under the mattress at the level of the knees. This will not only flex his knees but will keep him more comfortably in the upright position (Fig. 11.15).

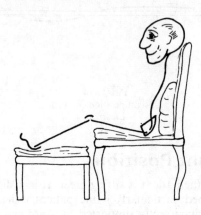

FIG. 11.16
Upright position—sitting.
Correct

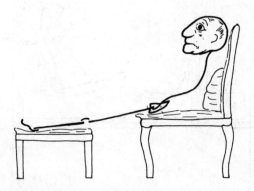

FIG. 11.17
Upright position—sitting.
Incorrect

If the patient is sitting in a chair, you must see that his spine is held upright and that the lumbar curve is supported by the use of pillows (Fig. 11.16). If a footstool is used, it should be only slightly lower than the chair and continuous with the seat of the chair so that the knees are not hyper-extended (Fig. 11.17). If properly supported, the patient will be comfortable because he is well balanced and no strain is being put on any of his ligaments and muscles.

Index

Abdomen, 5, 6
Acetabulum, 3
Acromegaly, 121
Adenoids, 72
Adipose tissue, 3
Adrenaline, 86, 124
Air passages, 69
Air sacs, 73
Albumen, 59, 62
Alimentary canal, 78
Alveoli, 73
Amino acids, 59, 77, 87, 90
Anatomical position, 13
Anatomy, 1
Antibodies, 59, 86
Anus, 83
Aorta, 62
Aperient, 84
Appendix, 82
Aqueous humour, 114
Arachnoid mater, 110
Arches (of foot), 38
Arteriosclerosis, 63
Arteries, brachial, 56
  carotid, 57
  coronary, 56
  facial, 57
  femoral, 57
  iliac, 57
  popliteal, 57
  pulmonary, 59
  radial, 56
  sub-clavian, 56
  temporal, 57
  tibial, 57
  ulnar, 56
Arterioles, 55
Articulations, 42
Arthritis, 44

Atlas, 17
Atria, 53, 62
Aural ossicles, 115
Axilla, 26
Axis, 17, 18

Bacillus Coli, 84
Bacteria, 60
Balance, 116
Bicuspid valve, 55
Bile, 82, 85
Biology, 1
Bladder, 101
Blood, 46, 53, 59
  cells, 60
  corpuscles, red, 60
  corpuscles, white, 60
  groups, 61
  oxygenated, 62
  pressure, 63, 65
    diastolic, 65
    systolic, 65
  venous, 62
  vessels, 53, 55
Bone, 2
  marrow, 60
    yellow, 13
  tissue, 11
    cancellous, 11
    spongy, 11
Bones, 9, 10
  flat, 10
  irregular, 10
  long, 9
  short, 9
Bones, See under Individual names
Brain, 105
Brain stem, 107
Bronchi, 73

Bronchial tubes, 73
Bronchioles, 73
Bursa, 34, 35
Buttock, 50

Caecum, 82
Cauda equina, 110
Calcaneum, 38
Calcium, 12, 59, 88, 123
Calories, 90
Capillaries, 55, 62
Carbohydrate, 59, 87, 89
Carbon dioxide, 47, 59, 62, 74, 90
Cardiac cycle, 64
Carpus, 30, 31
Cartilage, 3
Cell, 1
Cellulose, 77
Cerebellum, 108
Cerebral cortex, 106
Cerebral haemorrhage, 108
Cerebrospinal fluid, 106, 110
Cerebrum, 106
Cervix, 129
Choroid, 112
Chyme, 80
Ciliated epithelium, 2
Circulation, 61
Circulation, coronary, 56
  general, 56
  portal, 59, 85
  pulmonary, 59, 73
Circulatory system, 4, 53
Clavicle, 26
Clot, 59, 60, 86
Coccyx, 19
Colon, 82, 83
Conjunctiva, 113
Connective tissue, 2
Constipation, 84
Cornea, 111
Cortisone, 124
Costal cartilages, 25
Cranium, 5, 20
Cretin, 122

Defaecation, 84
Dermis, 97
Diabetes, 84, 90
Diaphram, 6, 48, 74
Diarrhoea, 84
Diastole, 65
Digestion, 91
Digestive system, 4, 77
Discs, 20
Drum, 115

Duodenum, 82
Dura mater, 110
Dwarfism, 121

Ear, 115
Embolism, 56
Endocardium, 54
Endocrine system, 4
Endometrium, 129
Energy, 46, 87
Enzymes, 77
Epidermis, 96
Epiglottis, 71
Epithelium, 2
Ergosterol, 98
Erythrocytes, 60
Ethmoid, 22
Excretory system, 4, 95
Exophthalmic goitre, 122
Expiration, 69
Eye, 111

Face, 22
Faeces, 84
Fat, 59, 90
Fats, 87
Fatty acids, 77
Fatty tissue, 2, 13
Femur, 33, 34
Fenestra ovale, 115
Fibrous tissue, 3, 11, 12
Fibula, 36
Foetus, 127
Fontanelles, 23
Food, 86
Foot, 37
Foramen magnum, 21
Fore arm, 29
Frontal bone, 20
Fundus, 128

Gall bladder, 85
Gastric juice, 80
Genitals, external, 129
Giant, 121
Girdle, pelvic, 31
  shoulder, 26
Glands, adrenal, 86, 123
  ductless, 119
  endocrine, 119
  gastric, 80
  intestinal, 81
  lachrymal, 114
  mammary, 129
  parathyroid, 123
  parotid, 79

pineal, 125
pituitary, 119
prostate, 131
salivary, 79
sebaceous, 98
sub-lingual, 79
sub-mandibular, 79
suprarenal, 123
sweat, 97
thymus, 125
thyroid, 121
Glans penis, 131
Glucose, 46, 59, 77, 86, 87, 90
Glycerin, 77
Glycogen, 86
Grey matter, 105

Haemoglobin, 60
Haemorrhage, 60
Hair, 97
Hand, 30
Hearing, 116
Heart, 53
Hip, 34
Hormones, 59, 119
Humerus, 27, 28
Hyaline cartilage, 3, 11, 12
Hyperthyroidism, 122

Ilium, 32
Iodine, 88
Incontinence, 102
Infection, 60, 66
Ingestion, 91
Innominate bone, 32
Inspiration, 69
Intestinal juice, 81
Intestine, large, 82
 small, 81
Insulin, 80, 84, 90, 125
Iris, 112
Iron, 60, 88
Ischium, 34
Islets of Langerhans, 84, 125

Jaundice, 61
Joints, 42
 capsule, 44
 freely movable, 42
 immovable, 42
 movements, 43
 slightly movable, 42
 synovial, 42
 types, 44

Ketones, 90

Kidneys, 99

Labia, 120
Labyrinth, bony, 115
Labyrinth, membraneous, 115
Lachrymal bones, 22
Lacteal, 82
Lactiferous tubules, 129
Larynx, 72
Lens, 113
Leucocytes, 60
Ligaments, 3
Limbs, 6
 lower, 34
 upper, 27
Liver, 82, 85, 90
Lungs, 69, 73
Lymph, 65
Lymphatic capillaries, 62, 65
 ducts, 65
 nodes, 65
 system, 65
 vessels, 65
Lymphocytes, 60, 65

Malar bones, 22
Malleolus, 36, 37
Mandible, 22
Marrow, 13
Maxillary bones, 22
Mediastinum, 67
Medulla oblongata, 108
Meninges, 110
Menstrual cycle, 129
Mesentery, 67, 81
Metabolism, 89
Metacarpals, 30, 31
Metatarsals, 37, 38
Micturition, 101
Midline, 13
Mineral salts, 88
Mitral valve, 55
Mouth, 5, 78
Mucous membrane, 2
 ciliated, 69
Muscles, 45
 abdominal, 48
 action, 46
 biceps, 49
 deltoid, 49
 diaphragm, 48
 gluteal, 50
 intercostals, 48, 74
 pelvic, 49
 quadriceps, 51
 structure, 45

thoracic, 48
trunk, 48
Muscular system, 4, 45
Muscle tissue, 2, 3
Muscle tissue, cardiac, 3
  involuntary, 3
  skeletal, 3
  voluntary, 3
Myocardium, 54
Myxoedema, 122

Nails, 97
Nasal bones, 22
Nephron, 100
Nerves, autonomic, 110
  cranial, 108
  lateral popliteal, 36
  of hearing, 116
  of smell, 71
  of taste, 80
  of touch, 96
  optic, 113, 114
  para-sympathetic, 111
  radial, 28
  sciatic, 33, 50
  spinal, 108
  sympathetic, 111
  ulnar, 29
Nervous system, 4, 104
Nervous tissue, 2, 3
Neural arch, 15
Neural canal, 15
Neurone, 105
Nose, 5, 70
Nucleus, 1

Occipital bone, 21
Oedema, 55, 62, 102
Oesophagus, 80
Oestrin, 128
Omentum, 86
Orbits, 5
Organs, 3, 4
Ossification, 12
Ova, 127
Ovaries, 125, 128
Oxygen, 2, 46, 60, 62, 69

Palatine bones, 23
Pancreas, 82, 84, 125
Pancreatic juice, 82, 84
Paralysis, 48
Paraplegia, 20
Parietal bones, 21
Patella, 35
Pelvis, 5, 6, 31

Penis, 131
Pericardium, 54
Perineum, 49
Periosteum, 12
Peristalsis, 80
Peritoneum, 86, 129
Phalanges, 30, 31, 37, 39
Pharynx, 50, 71
Pia mater, 110
Pituitrin, 121
Plasma, 59, 62
Platelets, 60
Pleura, 74
Popliteal space, 35
Position, lateral, 140
  prone, 138
  recumbent, 33
  supine, 137
  upright, 141
Posture, 133
Potassium, 59, 88
Prepuce, 131
Progesterone, 128
Protein, 59, 87, 90
Protoplasm, 1, 2
Puberty, 129, 130
Pubis, 34
Pulse, 56, 65
Pupil, 112
Pylorus, 80

Radius, 29
Rectum, 49, 83
Red blood corpuscles, 60
Red bone marrow, 13
Reflex action, 109
Reproduction, 129
Reproductive system, 4, 127
Renal calculi, 102
Respiration, 69, 74
Respiratory system, 4, 69
Retention, 102
Retina, 113
Rhesus factor, 61
Rickets, 12, 89
Ribs, 25
Roughage, 87

Sacrum, 19
Saliva, 79
Salts, 59, 123
Scapula, 26
Sclera, 111
Scrotum, 130
Sebum, 98
Semen, 131

Seminal vesicles, 131
Septum, 62
Serous membrane, 74, 81
Sesamoid bone, 35
Shoulder girdle, 26
Sigmoid colon, 82
Sinuses, 23, 71
Sinusitis, 24
Skeletal system, 4, 9
Skeleton, 9
Skin, 95
Skull, 20
Sodium chloride, 59, 88
Sphenoid bone, 22
Sphygmomanometer, 63
Spermatozoon, 129, 130
Spleen, 60, 67
Spinal cord, 108
Spinal curves, 16
Spine, 13
Starch, 87
Sternum, 24
Steroids, 123
Stomach, 80
Stroke, 63
Subarachnoid space, 110
Sugar, 87
Suppression of urine, 102
Sutures, 23
Sweat, 97
Synovial membrane, 43
Systems, 3–4
  circulatory, 53
  endocrine, 119
  excretory, 95
  digestive, 77
  muscular, 42
  nervous, 104
  reproductive, 127
  respiratory, 69
  skeletal, 9
Systole, 65

Talus, 38
Tarsus, 37
Tears, 114
Teeth, 78
Temporal bones, 21, 115
Tendon, 2, 45
Testes, 126, 130
Testosterone, 130
Thoracic duct, 67
Thorax, 5, 6, 24
Throat, 71
Thrombosis, 35, 56, 60
Thyroid cartilage, 72

Thyroxine, 122
Tibia, 36
Tissue, 2
Tissue fluid, 62
Tongue, 80
Tonsils, 67
Trachea, 72
Tricuspid valve, 55
Trunk, 5
Turbinate bones, 22

Ulna, 30
Urea, 59, 86, 90
Ureters, 101
Urethra, 49, 102, 131
Urinary system, 98
Urine, 101
Uterine tubes, 128
Uterus, 128

Vagina, 49, 129
Vasa deferentia, 131
Veins, 55
  jugular, 57
  portal, 59, 86
  pulmonary, 59, 62
  superficial, 58
  valves, 56
Vena cava, 57, 58, 62
Ventricles, 53
Venules, 55, 106
Vertebrae, 13
  cervical, 17
  coccygeal, 17
  curves, 16
  dorsal, 17, 18
  lumbar, 17, 19
  sacral, 17
  thoracic, 17, 18
Villi, 82
Vitreous humour, 114
Vitamins, 88
  A, 88
  B, 88
  B_{12}, 60, 80, 88
  C, 89
  D, 12, 89, 98
  E, 89
  K, 89
Vocal cords, 72
Vomer, 22
Vulva, 129

Water, 87
Water vapour, 47
White blood corpuscles, 60, 65
White matter, 105

Printed in Great Britain by
T. & A. Constable Ltd., Edinburgh